McGRAW-HILL RYERSON MATHEMATICS 8

MAKING CONNECTIONS
Student Workbook

ADVISORS

Dan Antflyck
Toronto District School Board

Richard Chaplinsky
Ottawa-Carleton Catholic District School Board

Warren Dixon
District School Board of Niagara

Christina Maschas-Hammond
Peterborough, Victoria, Northumberland, Clarington Catholic District School Board

Roland W. Meisel
Port Colborne, Ontario

Christopher Perry
Hamilton Wentworth District School Board

McGraw-Hill Ryerson

Toronto   Montréal   Boston   Burr Ridge, IL   Dubuque, IA   Madison, WI   New York
San Francisco   St. Louis   Bangkok   Bogotá   Caracas   Kuala Lumpur   Lisbon   London
Madrid   Mexico City   Milan   New Delhi   Santiago   Seoul   Singapore   Sydney   Taipei

COPIES OF THIS BOOK
MAY BE OBTAINED BY
CONTACTING:

McGraw-Hill Ryerson Ltd.

WEB SITE:
http://www.mcgrawhill.ca

E-MAIL:
orders@mcgrawhill.ca

TOLL-FREE FAX:
1-800-463-5885

TOLL-FREE CALL:
1-800-565-5758

OR BY MAILING YOUR
ORDER TO:
McGraw-Hill Ryerson
Order Department
300 Water Street
Whitby, ON L1N 9B6

Please quote the ISBN
and title when placing
your order.

Student Workbook ISBN:
0-07-091695-0

Mathematics 8 Making Connections Student Workbook

Copyright © 2005, McGraw-Hill Ryerson Limited, a Subsidiary of The McGraw-Hill Companies. All rights reserved. No part of this publication may be reproduced or transmitted in any form or by any means, or stored in a data base or retrieval system, without the prior written permission of McGraw-Hill Ryerson Limited, or, in the case of photocopying or other reprographic copying, a licence from The Canadian Copyright Licensing Agency (Access Copyright). For an Access Copyright licence, visit *www.accesscopyright.ca* or call toll free to 1-800-893-5777.

ISBN 0-07-091695-0

http://www.mcgrawhill.ca

2 3 4 5 6 7 8 9 10 M 0 9 8 7 6 5

Printed and bound in Canada

Care has been taken to trace ownership of copyright material contained in this text. The publishers will gladly accept any information that will enable them to rectify any reference or credit in subsequent printings.

National Library of Canada Cataloging in Publication Data

Mathematics 8: making connections. Student workbook.

ISBN 0-07-091695-0

1. Mathematics —Problems, exercises, etc.

HQ755.8.W57 2004 Suppl. 2   306.874   C2004-902813-8

PUBLISHER: Diane Wyman
WRITING/DEVELOPMENT: Eileen Jung, Bradley T. Smith, Wendi Morrison, Johnny Lam,
            and Laurel Sparrow of First Folio Resource Group, Inc.
MANAGER, EDITORIAL SERVICES: Linda Allison
SENIOR SUPERVISING EDITOR: Crystal Shortt
EDITORIAL ASSISTANT: Erin Hartley
PRODUCTION SUPERVISOR: Yolanda Pigden
PRODUCTION COORDINATOR: Jennifer Wilkie
ELECTRONIC PAGE MAKE-UP: Tom Dart/First Folio Resource Group, Inc.

Contents

Problem Solving v

Get Ready for Grade 8 x

Chapter 1 Measurement and Number Sense 1

Chapter 2 Two-Dimensional Geometry 11

Chapter 3 Fraction Operations 19

Chapter 4 Probability 33

Chapter 5 Ratio, Rates, and Percents 41

Chapter 6 Patterning and Algebra 53

Chapter 7 Exponents 61

Chapter 8 Three-Dimensional Geometry and Measurement 71

Chapter 9 Data Management: Collection and Display 81

Chapter 10 Data Analysis: Analysis and Interpretation 89

Chapter 11 Integers 101

Chapter 12 Patterning and Equations 119

Chapter 13 Geometry of Angle Properties 127

Preparing for Grade 9 137

Name: _____   Date: _____

Problem Solving

Understand

Read the problem.
- Think about the problem. Express it in your own words.
- What information do you have?
- What further information do you need?
- What is the problem asking you to do?

Plan

Select one or more strategies for solving the problem.
- Is this problem like one you solved before? Can you use a similar strategy?
  - Make a picture or diagram
  - Make an organized list
  - Look for a pattern
  - Make a model
  - Work backward
  - Make a table or chart
  - Act it out
  - Use systematic trial
  - Make an assumption
  - Find needed information
  - Choose a formula
  - Solve a simpler problem
- Decide whether any of the following might help. Plan how to use them.
  - tools such as a ruler or a calculator
  - materials such as graph paper or a number line

Do It!

Solve the problem by carrying out your plan.
- Use mental math to estimate a possible answer.
- Do the calculations and record your steps.
- Explain and justify your thinking.
- Revise your plan if it does not work out.

Look Back

Examine your answer. Does it make sense?
- Is your answer close to your estimate?
- Does your answer fit the facts given?
- Is the answer reasonable? If not, make a new plan. Try a different strategy.
- Consider solving the problem a different way. Do you get the same answer?
- Compare your method with that of other students.

Study Skills
Create your own problems. Exchange them with a study partner. Compare the strategies you and your partner used to solve the problem. Which strategy was easier to use? Why? Write notes about which strategy may be more efficient to use.

Four problems are presented.

Complete the solutions for each problem by completing diagrams and filling in the blanks.

After you have completed the problems, try using a different strategy.

Name: _____ Date: _____

Problem 1

A tourist train engine has a mass of 7.6 t and is 28 m long. Full passenger cars each have a mass of about 3.4 t and are 22 m long. The engine can pull up to 30 t, including its own mass. How many cars can one engine pull and how long is the entire train?

Make an organized list

a) Complete the table.

Diagram	Total mass (t)	Total length (m)
engine	7.6	28
1 car + engine	11.0	50
2 cars + engine		
3 cars + engine		

b) When does the mass become greater than 30 t? _____

c) The engine can pull up to _____ passenger cars.

Use systematic trial

The total mass is 7.6 + 3.4 × the number of passenger cars.

a) Try 8 cars:
   7.6 + 3.4 × ____
   = 7.6 + ____
   = ____ (Too high)

   Try 7 cars:
   7.6 + 3.4 × ____
   = 7.6 + ____
   = ____ (____)

   Try 6 cars:
   7.6 + 3.4 × ____
   = 7.6 + ____
   = ____ (____)

b) The engine can pull up to _____ passenger cars.

Choose a formula

a) The total length is 28 + 22 × _____.

b) What is the length of the longest possible train?
   28 + 22 × ____
   = 28 + ____
   = ____

c) The length of the train is _____.

vi   MHR • Problem Solving

Name: _____   Date: _____

Problem 2

A knight is riding back to the castle. At each bridge he crosses, he must pay a toll of one-third of his gold. After crossing 3 bridges, he has 8 gold coins left. How many gold coins did he start with?

Use systematic trial

a) How many gold coins did he have before the last bridge?

Try 9 coins:
$9 - 9 \div 3$
$= 9 -$ _____
$=$ _____ (Too low)

Try 15 coins:
$15 - 15 \div 3$
$= 15 -$ _____
$=$ _____ (_____)

Try 12 coins:
$12 - 12 \div 3$
$= 12 -$ _____
$=$ _____ (_____)

b) The number of gold coins before the last bridge was _____.

Work backward

a) The knight gives one-third of his coins away. What fraction does he keep? _____

b) How is one-third related to this fraction? How can you calculate the number of gold coins the knight had before reaching a bridge?

Make a table or chart

a) Complete the table.

Bridge	Coins After Crossing Bridge	Coins Before Reaching Bridge
3	8	12
2	12	
1		

b) How many gold coins did the knight start with? _____

Hint
Calculate the number of coins before reaching a bridge using guess and test or using fractions.

Problem 3

This field needs to be fenced in and corn seeds need to be sown. Fencing costs $3.75/m and a $30-bag of seeds will cover 500 m². What will be the total cost?

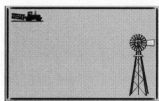

60 m

100 m

Choose a formula

a) The length of the field is _____.

   The width of the field is _____.

b) How do you calculate the perimeter of a rectangle?

c) The perimeter of the field is _____.

d) How do you calculate the area of a rectangle?

e) The area of the field is _____.

f) The cost of the fencing is 3.75 × the perimeter of the field.
   3.75 × _____ = _____

Make an assumption

Assume that seeds will be sown across the entire field.

a) How many bags of seed are needed?

b) The cost of the seeds is 30 × the number of bags needed.

   30 × _____ = _____

c) Add the cost of the fencing to the cost of the seeds.

d) The total cost is _____.

Hint

To calculate the number of bags needed, divide the field's area by the area one bag covers.

Name: _____   Date: _____

Problem 4

Some friends are having a movie marathon of 5 spooky films. Their only rule is they must watch *Zombies Awake* before *Zombies Exiled*. How many different orders are there to watch the movies?

Solve a simpler problem

If they only watch *Zombies Awake* (A) and *Zombies Exiled* (E) then there is only 1 possible ordering: AE.

a) If they also watch *Beastly Bats* (B), write out the possible ways to watch the 3 movies.

b) In how many ways can the 3 movies be viewed? _____

c) If there is also *The Creature is Coming* (C), write out the possible ways to watch the 4 movies.

d) In how many possible ways can the 4 movies be viewed? _____

Look for a pattern

a) For the arrangement BAEC, in how many ways can you also watch *Dracula's Dragons* (D) and keep the order of the other 4 the same? _____

b) For any arrangement of 4 movies, in how many ways can you also watch *Dracula's Dragons* and keep the order of the other 4 the same? _____

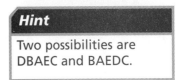

Hint

Two possibilities are DBAEC and BAEDC.

c) How many different orders are there in which to watch the movies? _____

1 Fractions, Decimals, SI Units, Estimation, Measurement

Get Ready Mentally

1. Is each answer greater than (>), less than (<), or equal to 1? How do you know?
   a) $0.5 + 0.05 + 0.005$
   b) 1.22
   c) $\frac{1}{3} + \frac{1}{6} + 0.5$
   d) $10.9 - 9.01$
   e) $-1.1 + 2.0$
   f) $\frac{4}{5} + \frac{1}{3} - \frac{1}{6}$

2. Is the perimeter greater than 30 m? Is the area greater than 600 m²? How do you know?

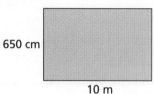

650 cm

10 m

Get Ready by Thinking

Choose the most reasonable estimate in each question. Explain your thinking.

3. The area of a computer's keyboard is about
   A 2 m^2
   B 700 cm^2
   C 0.5 m^2
   D 70 cm^2

4. The mass of a computer's mouse is about
   A 1.5 kg
   B 0.015 kg
   C 150 g
   D 15 000 g

5. The capacity of a box of tissues is about
   A 200 cm^3
   B $20\ 000 \text{ cm}^3$
   C 2 m^3
   D 2000 cm^3

6. The height of a basketball net is about
   A 0.1 km
   B 200 cm
   C 3 m
   D 3500 cm

7. The shaded portion is about
   A $\frac{5}{6}$
   B $\frac{2}{3}$
   C $\frac{1}{2}$
   D $\frac{3}{4}$

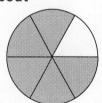

8. The time you spend asleep in one day is about what fraction of the day?
   A $\frac{2}{3}$
   B $\frac{1}{3}$
   C $\frac{11}{24}$
   D $\frac{1}{6}$

9. The volume of juice in a juice box is about
   A 2500 mL
   B 2.5 L
   C 0.25 L
   D 25 mL

10. Eight people share a $70 bill at a restaurant equally. About how much does each person pay? Estimate to the nearest dollar.
    A $7
    B $8
    C $9
    D $10

11. It costs $5.75 to rent a movie at a local store. How much change should you expect if you pay $20 for 3 movie rentals? Estimate to the nearest dollar.
    A $2
    B $3
    C $4
    D $5

x MHR • Get Ready for Grade 8

Name: _____   Date: _____

2 Convert Fractions, Decimals, and Percents; Perfect Squares and Square Roots

Get Ready Mentally

1. Write each fraction as a decimal and as a percent.
   a) $\frac{3}{10} =$
   $=$
   b) $\frac{3}{4} =$
   $=$
   c) $\frac{17}{20} =$
   $=$
   d) $\frac{18}{25} =$
   $=$

2. Which is greater? How do you know?
   a) $\frac{6}{8}$ or $\frac{4}{5}$
   b) $\frac{3}{10}$ or 0.35
   c) 0.2 or 0.08
   d) 0.682 or $\frac{2}{3}$

3. Write each decimal as a percent.
   a) 0.3 =
   b) 0.85 =
   c) 0.06 =
   d) 0.44 =
   e) 0.816 =
   f) 0.70 =

4. Which is greater? How do you know?
   a) 3^2 or 8
   b) 6^2 or 8^2
   c) 50 or 7^2
   d) 30^2 or 29^2

Get Ready by Thinking

5. What is the area of each square?
   a) 3 cm

   b) 5 m
   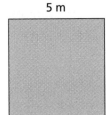
   c) 50 cm

   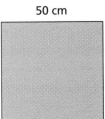
   d) 30 km

6. What is the side length of a square with each area?
   a) 25 m²
   b) 1 mm²
   c) 400 cm²
   d) 64 km²

7. What is the side length of a square with each perimeter?
   a) 36 cm
   b) 528 m
   c) 10 cm
   d) 5.6 km

8. What is the area of a square with each perimeter?
   a) 24 m
   b) 4 cm
   c) 40 km
   d) 44 mm

3 Patterns With Natural Numbers, Fractions, and Decimals

Get Ready Mentally

1. List the next three numbers in each pattern.
   a) −1, −4, −7, ____, ____, ____

   b) 224, 112, 56, ____, ____, ____

   c) 9, 5, 1, ____, ____, ____

   d) 1, 3, 6, 10, ____, ____, ____

   e) 10, 7, 12, 9, 14, 11, ____, ____, ____

   f) −5, 10, −20, 40, ____, ____, ____

2. List the next three numbers in each pattern.
   a) $\frac{4}{5}$, $1\frac{1}{5}$, $1\frac{3}{5}$, ____, ____, ____

   b) 2, $1\frac{1}{2}$, 1, ____, ____, ____

   c) $-\frac{3}{8}$, $-\frac{3}{4}$, $-\frac{3}{2}$, ____, ____, ____

   d) $\frac{5}{7}$, $1\frac{6}{7}$, 3, ____, ____, ____

Get Ready by Thinking

3. In each box, circle the number that does not belong.

a)
| 12 | −3 | 8 |
| −21 | 18 | −18 |

b)
| $\frac{4}{10}$ | $\frac{10}{25}$ | $\frac{6}{15}$ |
| $\frac{20}{50}$ | $\frac{2}{5}$ | $\frac{10}{30}$ |

c)
| 121 | 1 | 25 |
| 9 | 65 | 49 |

4. Describe what happens to the input number to get the output number.

a)
Input	Output
−3	1
0	4
$2\frac{1}{2}$	$6\frac{1}{2}$
6	10

b)
Input	Output
−3	−9
$-\frac{1}{3}$	−1
$\frac{1}{4}$	$\frac{3}{4}$
2	6

c)
Input	Output
−4	−15
0	−3
2	3
6	15

1.1 Discover the Pi Relationship
1.2 Circumference Relationships

Student Text pp. 12–21

Key Ideas Review

Fill in each blank with a letter, symbol, or word from the list. Many choices will be used twice.

C	r	d	$\doteq$
circumference	diameter	radius	π

1. Label the diagrams.

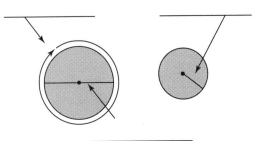

2. Write the short forms.

circumference _____

radius _____

diameter _____

approximately equal to _____

pi _____

3. Complete the formulas.

π _____ 3.14   _____ $= \dfrac{C}{d}$ _____ $= \pi \times d$ $d = 2 \times$ _____

Example 1: Diameter and Circumference Formula

Find the circumference of the circle. Round to the nearest centimetre.

Solution

$C = \pi \times d$
$C = \pi \times 14$
$C = 43.982...$
$C \doteq 44$

The circumference of the circle is about 44 cm.

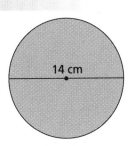

14 cm

Literacy Connections

Writing Answers
Always include proper units in the final answer to any measurement problem.

Name: _____   Date: _____

Example 2: Radius and Circumference Formula

Find the circumference of the circle. Round to the nearest kilometre.

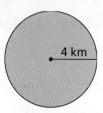

Solution

$C = 2 \times \pi \times r$

$C = 2 \times \pi \times 4$

$C = 25.132...$

$C \doteq 25$

The circumference of the circle is about 25 km.

Study Skills

Look for this feature throughout the book.

It will provide ideas to help you prepare for a chapter test.

Practise

1. Find the circumference of each circle. Round to the nearest metre.

a)

b)

2. Find the circumference of each circle. Round to the nearest centimetre.

a)

b)

Apply

3. Ryan's birthday cake is 22.5 cm in diameter. How many centimetres of icing are needed to trim the cake?

Making Connections

How can the formulas for area and circumference of a circle be used in baking?

In what other jobs would you apply the formulas?

2 MHR • Chapter 1: Measurement and Number Sense

1.3 Discover the Area of a Circle

Student Text pp. 22–25

Key Ideas Review

Circle the word or formula that will complete each sentence correctly.

1. The area of a circle is related to its **radius** / **centre**.

2. To calculate the area of a circle, use the formula **$A = \pi \times d^2$** / **$A = \pi \times r^2$**.

3. The formula **$3 \times r^2$** / **$3 \times \pi$** can be used to estimate the area of a circle.

Example 1: Calculate Area From Radius

Find the area of the circle. Round to the nearest tenth of a square metre.

Solution

$A = \pi \times r^2$
$A = \pi \times (1.3)^2$ $(1.3)^2 = 1.3 \times 1.3$
$A = \pi \times 1.69$
$A = 5.309...$
$A \doteq 5.3$

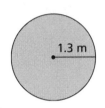

Literacy Connections

Reading Squares
Read r^2 as "r squared."
It means $r \times r$.

The area of the circle is about 5.3 m².

Example 2: Calculate Area From Diameter

Find the area of a circular swimming pool cover that measures 4 m across.

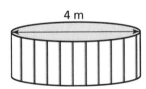

Solution

Find the radius.

$r = d \div 2$
$r = 4 \div 2$
$r = 2$

Apply the area formula.

$A = \pi \times r^2$
$A = \pi \times 2^2$
$A = \pi \times 4$ Estimate: $3 \times 4 = 12$
$A = 12.566...$
$A \doteq 12.6$

The area of the swimming pool cover is approximately 12.6 m².

Name: _____   Date: _____

Practise

1. Find the area of each circle. Round to the nearest tenth of a square kilometre.

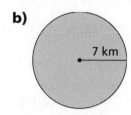

a) 2.1 km   b) 7 km

2. Find the area of each circle. Round to the nearest square millimetre.

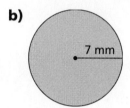

a) 116 mm   b) 7 mm

Apply

3. A lighthouse has two round floors that need to be painted. One measures 10 m across and the other measures 9 m across. What is the area that needs to be painted?

4 MHR • Chapter 1: Measurement and Number Sense

1.4
1.5

Draw Circles Using a Set of Compasses
Construct Circles From Given Data

Student Text pp. 26–32

Key Ideas Review

Fill in each blank with a word or phrase from the list.

| folds | centre point | radius | points |

1. To draw a circle using a set of compasses, you need to know the _____ or the _____ and an edge point.

2. You can also construct a circle that passes through three given points. You make two folds. Each one makes a pair of _____ that line up. The _____ goes where the folds cross.

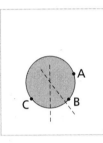

Practise

1. Draw a circle with each radius.

   a) 15 mm

   b) 2 cm

2. Construct a circle with each radius.

   a)

   b)

1.4 Draw Circles Using a Set of Compasses, 1.5 Construct Circles From Given Data • MHR 5

3. Draw a circle with centre A passing through point B.

a)

B.

A.

b)

.A

B.

4. Find the centre of the circle that passes through the three points shown. Draw the circle.

a)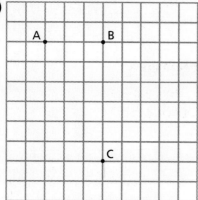

b)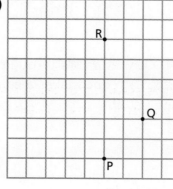

Making Connections

Create labels for the lids of various jars.

Explain to a friend how to do this.

Determine how much paper is needed for each label.

How many labels can you fit on a regular sheet of paper?

Decorate the jars with your labels.

Apply

5. Three friends are using a map to find a meeting place that is the same distance from each of their houses. Explain how they can find a solution using a circle.

1.6 Choose and Apply Circle Formulas

Student Text pp. 36–41

Key Ideas Review

Draw a line to match each statement in Column A with the best answer in Column B.

A	B
1. diameter	a) $2 \times \pi \times r$
2. circumference of a circle	b) $\pi \times r^2$
3. area of a circle	c) $3.14 \times d$
4. systematic trial	d) $2 \times r$
5. approximate circumference of a circle	e) guess and check

Example 1: Diameter and Radius From the Circumference

The 11.5 km bike path that goes around the lake is circular. How far apart are two cyclists when they are directly across the lake from each other?

Solution

$C = \pi \times d$ I know the circumference.
$11.5 = \pi \times d$ I need to find the diameter.

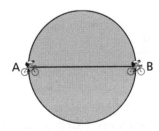

Use various possible values for the d variable.
Aim for an answer of about 11.5 km.

$\pi \times 3.8 = 11.938...$ Too high. The diameter should be about $11.5 \div 3$, which is a little less than $12 \div 3 = 4$.

$\pi \times 3.7 = 11.623...$ Close, but high.

$\pi \times 3.6 = 11.309...$ Close, but low.

To the nearest tenth of a kilometre, the closest value is $d = 3.7$.
The cyclists are about 3.7 km apart when they are directly across the lake from each other.

Literacy Connections

Reading Math Problems
What data are you given? What are you being asked to find? How can the data help you?

Example 2: Circumference From Radius.

A round mirror has a radius of 32 cm. What is the circumference of the frame?

Solution

$C = 2 \times \pi \times r$
$C = 2 \times \pi \times 32$
$C = 201.061...$
$C \doteq 201$

The circumference of the mirror is about 201 cm, or 2.01 m.

Practise

1. A line of skaters is 4 m long. One end of the line turns on the spot while the rest of the line skates in a circle around the fixed end, like a spinning needle on a compass. How far does the person on the moving end skate for one rotation, to the nearest metre?

2. Trudi is making a drum by tying a circle of leather over a juice can. The juice can is 15 cm in diameter. She needs the diameter of the leather circle to be 4 cm wider than the can. How much leather does she need, to the nearest square centimetre?

3. The ends of a 185 cm yellow ribbon just touch when it is stretched around an old oak tree. How thick is the tree, to the nearest centimetre?

Chapter 1: Reviewing for the Test

1.1 Discover the Pi Relationship
1.2 Circumference Relationships

page 12

1. Find the circumference of the circle. Round to the nearest centimetre.

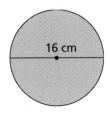

2. Find the circumference of the circle. Round to the nearest metre.

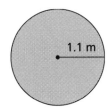

3. A circular jogging track is 50 m across the centre. How far will you jog if you go four times around the track?

1.3 Discover the Area of a Circle

page 22

4. Find the area of each circle. Round to the nearest square millimetre.

a)

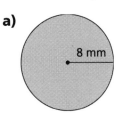

b)

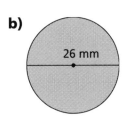

5. A circular tablecloth has a diameter of 2.4 m. How much material is needed to make it?

Name: _____   Date: _____

1.4 / 1.5 Draw Circles Using a Set of Compasses / Construct Circles From Given Data

page 26

6. Draw a circle with radius 17 mm. Explain your method.

7. Construct a circle with the given radius. Explain how you drew it.

8. Find the centre of the circle that passes through points T, U, and V. Draw the circle.

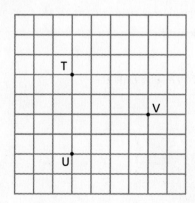

1.6 Choose and Apply Circle Formulas

page 36

9. An elastic can stretch to a maximum of 8.9 cm without breaking. What is the diameter of the largest rolled-up poster it can hold?

10. The circular garden in the front yard has a diameter of 2.7 m.

   a) How much will it cost to cover the garden in bark chips if each bag covers one square metre and costs $19.99?

   b) What length of garden fence is needed to enclose the garden?

Name: _____ Date: _____

2.1 Discover the Pythagorean Relationship

Student Text pp. 50–57

Key Ideas Review

Fill in each blank with a word, phrase, or number from the list. One choice will be used twice.

| 5 | leg squares | 25 | Pythagorean | right angle |
| 16 | hypotenuse | 3 | right triangle | 4   9 |

1. The _____ relationship tells how the areas of the _____ on the sides of a _____ are related.

2. The name for a shorter side of a right triangle is _____.

3. The side across from the _____ is called the _____.

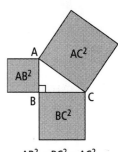

$AB^2 + BC^2 = AC^2$

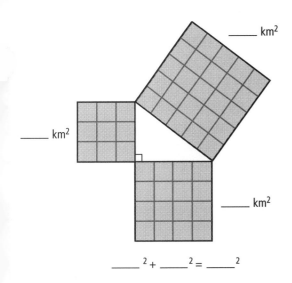

____^{2} + ____^{2} = ____^{2}

Example 1: Find the Area of the Square on the Hypotenuse

Find the area of the square on side AC.

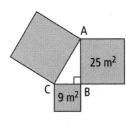

Solution

$3^2 + 5^2 = AC^2$
$9 + 25 = AC^2$
$34 = AC^2$

The area of the square on side AC is 34 m².

Name: _____ Date: _____

Example 2: Find the Area of the Square on a Leg of a Right Triangle

What is the area of the square on side PQ?

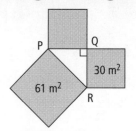

Solution

$PQ^2 + 30 = 61$
$ PQ^2 = 31$

The area of the square on side PQ is 31 cm².

Practise

1. What is the area of the square on the hypotenuse of each triangle?

   a)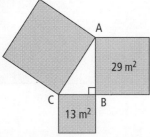

   b)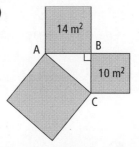

2. Find the area of the square on the third side of each triangle.

   a)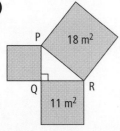

   b)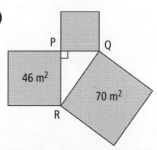

3. Identify the hypotenuse in each figure. Use the Pythagorean relationship to find the missing areas.

   a)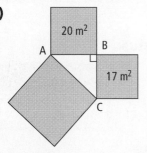

   b)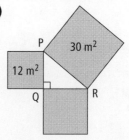

4. Identify the type of triangle. Then find the missing areas. Explain your answers.

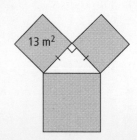

12 MHR • Chapter 2: Two-Dimensional Geometry

2.2 Find Approximate Values of Square Roots

Student Text pp. 58–61

Key Ideas Review

1. Draw a line to match each statement in Column A with the best answer in Column B.

   A

   1. Has a whole number for its square root
   2. A reasonable guess
   3. Not the square of a whole number
   4. Answer when an exact value is not possible

   B

   a) Estimate
   b) Non-perfect square
   c) Approximate value
   d) Perfect square

2. Circle the perfect squares.

   36   86   49   2.5   1.21   10   81   141   124   100

Example 1: Find Exact Square Roots

Evaluate.

a) $\sqrt{121}$
b) $\sqrt{0.16}$

Solution

a) $11 \times 11 = 121$, so $\sqrt{121} = 11$.
b) $0.4 \times 0.4 = 0.16$, so $\sqrt{0.16} = 0.4$.

Example 2: Find Approximate Square Roots

a) Estimate $\sqrt{7}$, to one decimal place.
b) Use a calculator to find the approximate value of $\sqrt{7}$. Round to two decimal places.

Solution

a) 4 and 9 are the two perfect squares closest to 7.
$2^2 = 4$ and $3^2 = 9$, so $\sqrt{7}$ is between 2 and 3.
7 is a little closer to 9 than it is to 4.
I estimate the value of $\sqrt{7}$, to one decimal place, to be 2.6.

Strategies
Make a picture or diagram.

b)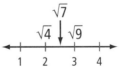
The value of $\sqrt{7} \doteq 2.65$.

Name: _____   Date: _____

Practise

1. Give the value of each square root.

   a) $\sqrt{3600}$

   b) $\sqrt{256}$

   c) $\sqrt{2.25}$

   d) $\sqrt{0.81}$

   Calculator Tip

   If your calculator doesn't have a $\sqrt{x}$ button, you may have to use the inverse key (INV) and the square key (x^2) because taking the square root of a number is the opposite of squaring a number.

2. Estimate each square root to one decimal place. Use a calculator to find the approximate value. Round to two decimal places.

   a) $\sqrt{60}$

   b) $\sqrt{43}$

   c) $\sqrt{110}$

   d) $\sqrt{6}$

   e) $\sqrt{15}$

   f) $\sqrt{99}$

Apply

3. Claire is making a square chalkboard. One small bottle of chalkboard paint covers 1.65 m². If she wants the chalkboard to be as large as possible, how many centimetres wide should she make the chalkboard? Explain your answer.

   Making Connections

   Examine a safety ramp or a skateboarding ramp.

   Measure the dimensions of the ramp.

   Identify the type of triangle a side of the ramp resembles.

   Explain how you know it is this type of triangle.

   Make a sketch of the ramp and include measurements.

Chapter 2: Reviewing for the Test

2.1 Discover the Pythagorean Relationship

page 50

1. For each figure:
   - identify the hypotenuse
   - find the missing area

   a)
   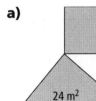

   b)
   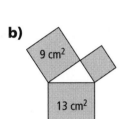

   c)
   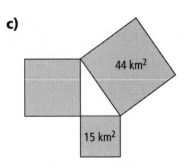

2.2 Find Approximate Values of Square Roots

page 58

2. Give the value of each square root.

   a) $\sqrt{1.96}$

   b) $\sqrt{144}$

   c) $\sqrt{16\,900}$

3. Estimate each value to one decimal place. Then, use a calculator to find the approximate value, to two decimal places.

   a) $\sqrt{54}$

   b) $\sqrt{136}$

4. Maggie used 8 m² of fabric to make a square quilt. What were the dimensions of the quilt?

2.3 Apply the Pythagorean Relationship
2.4 Use the Pythagorean Relationship to Solve Problems

page 62

5. Find the missing side length. Round to the nearest tenth, if necessary.

a)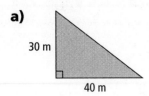

b)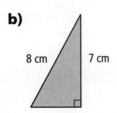

6. If the hypotenuse of a right triangle is 6 cm and one leg is 3 cm, find the length of the other leg. Round to two decimal places.

7. Which of the following could be lengths for a right triangle?

a) 4 cm, 5 cm, 6 cm

b) 1.2 m, 1.5 m, 0.9 m

8. Ravi cut a 21.6 cm by 28 cm sheet of paper along the diagonal to make two triangles. What was the perimeter of each triangle?

9. To get from school to the library, you can walk two blocks north and one block west, or you can take the diagonal path through the field. How much shorter is it if you take the path?

3.1 Add and Subtract Fractions

Student Text pp. 82–87

Key Ideas Review

Fill in each blank with a phrase or number from the list. One number may be used more than once.

| LCM | 2 | 12 | equivalent fractions | LCD |

1. The LCD is the _____ of the denominators of two or more fractions.

2. To add $\frac{2}{3}$ and $\frac{3}{4}$, use the _____.

3. The LCD of $\frac{2}{3}$ and $\frac{3}{4}$ is _____.

4. $\frac{2}{3} = \frac{4 \times}{4 \times 3} = \frac{8}{\_\_\_}$

   _____

Example 1: Add Fractions

Add $\frac{1}{2} + \frac{1}{3}$.

Solution

Method 1: Use Concrete Models (Pattern Blocks)
A double hexagon is equal to 1.

$\frac{1}{2} + \frac{1}{3} = \frac{5}{6}$

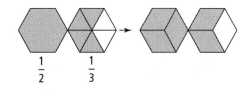

$\frac{1}{2}$ $\frac{1}{3}$

Method 2: Use Multiples to Find the Lowest Common Denominator (LCD)

The denominator of $\frac{1}{2}$ is 2.   The denominator of $\frac{1}{3}$ is 3.

Multiples of 2 are 2, 4, ⑥, 8, …   Multiples of 3 are 3, ⑥, 9, 12, …

The first multiple in both lists is 6.   The LCM of 2 and 3 is 6. The LCD of $\frac{1}{2}$ and $\frac{1}{3}$ is 6.

Write equivalent fractions with 6 as the denominator.

$\frac{1}{2} = \frac{1 \times 3}{2 \times 3}$ $\frac{1}{3} = \frac{1 \times 2}{3 \times 2}$

$= \frac{3}{6}$ $= \frac{2}{6}$

To find equivalent fractions, multiply the numerator and the denominator by the same number.

$\frac{1}{2} + \frac{1}{3} = \frac{3}{6} + \frac{2}{6}$

$= \frac{3 + 2}{6}$ Add the numerators.

$= \frac{5}{6}$

> **lowest common multiple (LCM)**
> - the lowest common multiple that two or more numbers have in common
> - the LCM of 4 and 5 is 20

Example 2: Subtract Fractions Using Factor Trees to Find the LCD

Subtract $\frac{4}{5} - \frac{1}{4}$.

Solution

The denominator of $\frac{4}{5}$ is 5.   The denominator of $\frac{1}{4}$ is 4.

Use a factor tree to write each denominator as a product of its prime factors.

$5 = 1 \times 5$ $4 = 2 \times 2$

To find the LCD, calculate the least number with all the prime factors of each denominator.

The LCD is $2 \times 2 \times 5 = 20$.

$\frac{4}{5} = \frac{4 \times 4}{5 \times 4}$ $\frac{1}{4} = \frac{1 \times 5}{4 \times 5}$

$= \frac{16}{20}$ $= \frac{5}{20}$

Write equivalent fractions with 20 as the denominator.

$\frac{4}{5} - \frac{1}{4} = \frac{16}{20} - \frac{5}{20}$

$= \frac{11}{20}$

prime factor
- a number with exactly two different factors, 1 and itself
- 3 is a prime number with factors 1 and 3

Practise

1. Write an addition sentence to represent the fraction of each figure shaded.

a)

_____ + _____ = _____

b)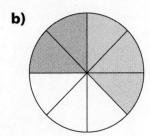

_____ + _____ = _____

2. Write a subtraction sentence to represent the fraction of each figure that remains.

a) 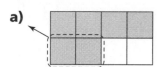   _____ − _____ = _____

b)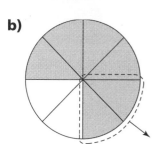

_____ − _____ = _____

3. From its prime factors, find the LCD for each set of fractions.

a) $\frac{3}{5}$ and $\frac{7}{8}$ factors of 5: _____

   factors of 8: _____

   LCD = _____

b) $\frac{4}{9}$ and $\frac{5}{12}$ factors of 9: _____

   factors of 12: _____

   LCD = _____

4. Add or subtract.

a) $\frac{1}{3} + \frac{1}{4} =$ _____ + _____

   $=$ $\frac{\_\_\_ + \_\_\_}{\_\_\_}$

   $=$ _____

b) $\frac{5}{6} + \frac{1}{3} =$ _____ + _____

   $=$ $\frac{\_\_\_ + \_\_\_}{\_\_\_}$

   $=$ _____

c) $\frac{5}{12} - \frac{5}{18} =$ _____ + _____

   $=$ $\frac{\_\_\_ + \_\_\_}{\_\_\_}$

   $=$ _____

3.2 Investigate Multiplying Fractions

Student Text pp. 88–91

Key Ideas Review

Fill in each blank with a word, number, or mathematical operator from the list.

| 4 | × | numerators | denominators |

1. $\dfrac{1}{\_\_}$ of 4 means the same as $\dfrac{1}{4}$ ____ 4.

   ____

2. To multiply fractions, from top to bottom, multiply the _____ together and then multiply the _____ together.

Example: Multiply Fractions

Multiply $\dfrac{1}{3} \times \dfrac{2}{4}$.

Solution

Method 1: Draw a Diagram

First, draw $\dfrac{1}{3}$. Then, draw $\dfrac{2}{4}$ of that.

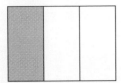 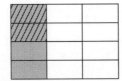

There are 12 parts in all. There are 2 parts that are striped.

So, $\dfrac{1}{3} \times \dfrac{2}{4} = \dfrac{2}{12}$ or $\dfrac{1}{6}$.

Method 2: Multiply Numerators and Denominators

$\dfrac{1}{3} \times \dfrac{2}{4} = \dfrac{\text{parts in product}}{\text{parts in whole}}$

$= \dfrac{1 \times 2}{3 \times 4}$

$= \dfrac{2}{12}$

$= \dfrac{1}{6}$

Name: _____   Date: _____

Practise

1. Rewrite each statement as a multiplication.

   a) $\frac{1}{5}$ of 8 = _____
   b) $\frac{1}{3}$ of 3 = _____
   c) $\frac{1}{6}$ of 10 = _____
   d) $\frac{1}{7}$ of 14 = _____

2. Multiply and write in lowest terms.

   a) $\frac{1}{2} \times \frac{4}{5} = \frac{\_\_\times\_\_}{\_\_\times\_\_} = \frac{\_\_}{\_\_}$
   b) $\frac{2}{7} \times \frac{2}{3} = \frac{\_\_\times\_\_}{\_\_\times\_\_} = \frac{\_\_}{\_\_}$
   c) $\frac{1}{2} \times \frac{1}{2} = \frac{\_\_\times\_\_}{\_\_\times\_\_} = \frac{\_\_}{\_\_}$
   d) $\frac{1}{3} \times \frac{5}{8} = \frac{\_\_\times\_\_}{\_\_\times\_\_} = \frac{\_\_}{\_\_}$
   e) $\frac{6}{5} \times \frac{5}{8} = \frac{\_\_\times\_\_}{\_\_\times\_\_} = \frac{\_\_}{\_\_}$
   f) $\frac{3}{6} \times \frac{5}{7} = \frac{\_\_\times\_\_}{\_\_\times\_\_} = \frac{\_\_}{\_\_}$

3. Calculate.

   a) $\frac{1}{5} \times 5 = $ _____
   b) $\frac{6}{7} \times 7 = $ _____
   c) $\frac{1}{4} \times 8 = $ _____
   d) $\frac{1}{3} \times 9 = $ _____
   e) $\frac{1}{2} \times 9 = $ _____
   f) $\frac{1}{3} \times 8 = $ _____

4. Find each amount.

   a) $\frac{1}{6}$ of $12 = $ ____ × $____
       = $____
   b) $\frac{2}{5}$ of $15 = $ ____ × $____
       = $____
   c) $\frac{5}{9}$ of $63 = $ ____ × $63
       = $____
   d) $\frac{3}{8}$ of $48 = $ ____ × $____
       = $____

5. $\frac{1}{4}$ of the students in a class have green eyes. $\frac{1}{3}$ of these students have blond hair. What fraction of the students have green eyes and blond hair? _____

6. Terrence has $15. He decides to spend $\frac{2}{5}$ on lunch and $\frac{1}{5}$ on a magazine.

   How much does he spend on lunch? $_____

   How much does he spend on the magazine? $_____

3.3 Investigate Dividing Fractions

Student Text pp. 92–95

Key Ideas Review

Fill in each blank with a word, phrase, or number from the list.

| 1 | reciprocals | $\frac{3}{2}$ | divide | multiply |

1. The product of two numbers that are _____ of each other is _____.

2. _____ is the reciprocal of $\frac{2}{3}$.

3. To _____ two fractions, write the reciprocal of the divisor (invert) and _____.

Example: Find Each Quotient

a) $4 \div \frac{1}{2}$ b) $\frac{1}{2} \div 4$ c) $\frac{1}{3} \div \frac{1}{2}$

Solution

Method 1: Draw a Diagram

a) $4 \div \frac{1}{2}$

The question is asking how many groups of $\frac{1}{2}$ can be made from 4.

Draw 4 circles. Each circle represents 1.   Divide each circle in half.

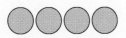   

There are 8 half-circles.

$4 \div \frac{1}{2} = 8$

b) $\frac{1}{2} \div 4$

The question is asking how many groups of 4 can be made from $\frac{1}{2}$.

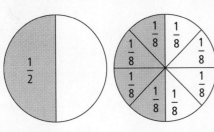

Divide the $\frac{1}{2}$ into 4 parts.

One of the 4 shaded parts represents $\frac{1}{8}$ of a whole.

$\frac{1}{2} \div 4 = \frac{1}{8}$

Name: _____ Date: _____

Method 2: Multiply By the Reciprocals

a) $4 \div \dfrac{1}{2} = 4 \times \dfrac{2}{1}$

 $= \dfrac{4 \times 2}{1}$

 $= 8$

b) $\dfrac{1}{2} \div 4 = \dfrac{1}{2} \times \dfrac{1}{4}$

 $= \dfrac{1 \times 2}{2 \times 8}$

 $= \dfrac{2}{16}$

 $= \dfrac{1}{8}$

c) $\dfrac{1}{3} \div \dfrac{1}{2} = \dfrac{1}{3} \times \dfrac{2}{1}$

 $= \dfrac{1 \times 2}{3 \times 1}$

 $= \dfrac{2}{3}$

Practise

1. Find each quotient.

a) $4 \div \dfrac{1}{4} =$ _____

b) $7 \div \dfrac{3}{2} =$ _____

c) $\dfrac{1}{3} \div 4 =$ _____

2. Divide.

a) $\dfrac{2}{3} \div \dfrac{5}{6} = \dfrac{}{} \times \dfrac{}{}$

 $= \dfrac{ \times }{ \times }$

 $= \dfrac{}{}$

 $= \dfrac{}{}$

b) $\dfrac{2}{5} \div \dfrac{8}{10} = \dfrac{}{} \times \dfrac{}{}$

 $= \dfrac{ \times }{ \times }$

 $= \dfrac{}{}$

 $= \dfrac{}{}$

c) $\dfrac{9}{10} \div \dfrac{1}{5} = \dfrac{}{} \times \dfrac{}{}$

 $= \dfrac{ \times }{ \times }$

 $= \dfrac{}{}$

 $= \dfrac{}{}$

3. Write each missing value.

a) $\dfrac{2}{5} \times \underline{} = \dfrac{4}{5}$

b) $\underline{} \times \dfrac{3}{7} = \dfrac{6}{7}$

c) $\dfrac{1}{3} \div \underline{} = 3$

d) $\dfrac{3}{5} \div \underline{} = \dfrac{3}{2}$

4. A theatre presentation consists of a series of plays. Each play is $\dfrac{2}{5}$ h. The entire presentation lasts $1\dfrac{1}{2}$ h. There is one intermission, which is less than $\dfrac{4}{10}$ h.

a) How many plays are performed? _____

b) How long is the intermission? _____

3.4 Order of Operations With Fractions

Student Text pp. 96–99

Key Ideas Review

Fill in each blank with a word or phrase from the list to show what operation is done in each step.

> exponent multiply convert to a mixed number
> do operation in brackets add reduce to lowest terms
> convert to equivalent fractions

$\left(\frac{1}{2}+\frac{1}{3}\right)^2 + \frac{2}{3} \times \frac{5}{6} = \left(\frac{5}{6}\right)^2 + \frac{2}{3} \times \frac{5}{6}$ 1. _____

$= \frac{25}{36} + \frac{2}{3} \times \frac{5}{6}$ 2. _____

$= \frac{25}{36} + \frac{10}{18}$ 3. _____

$= \frac{25}{36} + \frac{20}{36}$ 4. _____

$= \frac{45}{36}$ 5. _____

$= 1\frac{9}{36}$ 6. _____

$= 1\frac{1}{4}$ 7. _____

Example: Evaluate Expressions With Fractions and Brackets

Use the order of operations to evaluate $\left(\frac{4}{7} - \frac{2}{5}\right) \times \frac{1}{2}$.

Solution

$\left(\frac{4}{7} - \frac{2}{5}\right) \times \frac{1}{2} = \left(\frac{20}{35} - \frac{14}{35}\right) \times \frac{1}{2}$ Do the operation within brackets first. Change the fractions to equivalent fractions.

$= \frac{6}{35} \times \frac{1}{2}$ Now multiply.

$= \frac{6}{70}$ Reduce to lowest terms.

$= \frac{3}{35}$

Name: _____   Date: _____

Practise

Express all answers in lowest terms.

1. Circle the operation you would do first to evaluate the expression.

   a) $\dfrac{1}{4} + \dfrac{1}{2} \div \dfrac{1}{4}$
   b) $\dfrac{3}{8} + \dfrac{3}{4} \div \dfrac{2}{3}$
   c) $\dfrac{1}{4} \times \dfrac{1}{2} - \dfrac{1}{8}$

   d) $\dfrac{9}{10} + \dfrac{1}{3} \div \dfrac{1}{2}$
   e) $\left(\dfrac{3}{8} + \dfrac{3}{4}\right) \div \dfrac{2}{3}$
   f) $\left(\dfrac{1}{2}\right)^2 \times \dfrac{1}{4}$

2. Evaluate.

   a) $\dfrac{3}{7} + \dfrac{1}{21} - \dfrac{3}{14} = $ ____ $+$ ____ $-$ ____
   b) $\dfrac{4}{9} + \dfrac{5}{6} - \dfrac{2}{12} = $ ____ $+$ ____ $-$ ____

   $=$ ____

   $=$ ____

   $=$ ____

   $=$ ____

   $=$ ____

3. Evaluate.

   a) $\dfrac{3}{4} \times \dfrac{4}{8} - \dfrac{1}{12} = \dfrac{\times}{\times} - $ ____
   b) $\dfrac{8}{12} - \dfrac{1}{4} \times \dfrac{3}{5} = $ ____ $- \dfrac{\times}{\times}$

   $=$ ____ $-$ ____

   $=$ ____ $-$ ____

   $=$ ____ $-$ ____

   $=$ ____

   $=$ ____

4. Evaluate.

   a) $\left(\dfrac{8}{9} + \dfrac{2}{3}\right) \times \dfrac{3}{2} = $ ____ $\times$ ____
   b) $\dfrac{5}{2} \times \left(\dfrac{7}{8} - \dfrac{3}{4}\right) = $ ____ $\times$ ____

   $=$ ____

   $=$ ____

   $=$ ____

5. Insert brackets where necessary to make each equation true.

   a) $\dfrac{3}{4} \times \dfrac{5}{6} + \dfrac{1}{6} = \dfrac{3}{4}$
   b) $6 \times \dfrac{1}{3} + \dfrac{1}{2} = 5$

   c) $\dfrac{3}{4} \div \dfrac{3}{2} + \dfrac{1}{2} \times \dfrac{1}{2} = \dfrac{3}{16}$
   d) $\dfrac{1}{3} - \dfrac{1}{4} \times 12 + 2 = 3$

3.5 Operations With Mixed Numbers

Student Text pp. 100–105

Key Ideas Review

Fill in each blank with a word from the list.

> add   multiply   subtract   divide

1. When you _____ or _____ mixed numbers, you **do not** have to change them to improper fractions first.

2. When you _____ or _____ mixed numbers, you **do** have to change them to improper fractions first.

Example: Add, Subtract, and Multiply Mixed Numbers

Each day, Esther swims $2\frac{3}{4}$ lengths of front crawl, $1\frac{5}{8}$ lengths of butterfly, and $2\frac{1}{6}$ lengths of backstroke. She does this every day for 5 days.

a) In one day, how many more lengths of front crawl and butterfly than backstroke did she swim?
b) How many lengths of the front crawl does she complete in 5 days?

Solution

a)
$$2\frac{3}{4} = \frac{2 \times 4 + 3}{4} \qquad 1\frac{5}{8} = \frac{1 \times 8 + 5}{8} \qquad 2\frac{1}{6} = \frac{2 \times 6 + 1}{6}$$

$$= \frac{11}{4} \qquad\qquad\qquad = \frac{13}{8} \qquad\qquad\qquad = \frac{13}{6}$$

$$\frac{11}{4} = \frac{11 \times 6}{4 \times 6} \qquad \frac{13}{8} = \frac{13 \times 3}{8 \times 3} \qquad \frac{13}{6} = \frac{13 \times 4}{6 \times 4}$$

$$= \frac{66}{24} \qquad\qquad\qquad = \frac{39}{24} \qquad\qquad\qquad = \frac{52}{24}$$

Convert the mixed numbers to improper fractions. Convert these to equivalent fractions with the same denominator. The LCD is 24.

Now calculate.

$$2\frac{3}{4} + 1\frac{5}{8} - 2\frac{1}{6} = \frac{66}{24} + \frac{39}{24} - \frac{52}{24}$$

$$= \frac{66 + 39 - 52}{24}$$

$$= \frac{53}{24}$$

$$= 2\frac{5}{24}$$

To find how many more lengths of front crawl and butterfly Esther swam than of backstroke, add the $2\frac{3}{4}$ lengths of front crawl and the $1\frac{5}{8}$ lengths of butterfly. Then subtract the $2\frac{1}{6}$ lengths of backstroke.

28   MHR • Chapter 3: Fraction Operations

b) $2\frac{3}{4} \times 5 = \frac{11}{4} \times 5$ or $2\frac{3}{4} \times 5 = 2 \times 5 + \frac{3}{4} \times 5$

$\phantom{b)\ 2\frac{3}{4} \times 5\ } = \frac{55}{4}$ $\phantom{or\ 2\frac{3}{4} \times 5\ } = 10 + \frac{15}{4}$

$\phantom{b)\ 2\frac{3}{4} \times 5\ } = 13\frac{3}{4}$ $\phantom{or\ 2\frac{3}{4} \times 5\ } = 10 + 3\frac{3}{4}$

$\phantom{or\ 2\frac{3}{4} \times 5\ } = 13\frac{3}{4}$

To find how many lengths of the front crawl Esther swims in five days, multiply the number of lengths she swims in one day by 5.

Hint

When you multiply a mixed number, you can change it to an improper fraction first, but you don't have to. Either way, the answer is the same.

Esther swims $13\frac{3}{4}$ of the front crawl in one week.

Practise

Express all answers in lowest terms.

1. Change each mixed number to an improper fraction.

a) $4\frac{1}{2} =$ ──   b) $3\frac{1}{3} =$ ──   c) $4\frac{1}{8} =$ ──

2. Change each improper fraction to a mixed number.

a) $\frac{11}{2} =$ ──   b) $\frac{17}{5} =$ ──   c) $\frac{27}{10} =$ ──

3. Evaluate.

a) $6 - 1\frac{1}{3} =$ ── − ──   b) $3\frac{1}{3} - \frac{1}{6} =$ ── − ──

$\phantom{a)\ 6 - 1\frac{1}{3}} =$ ── $\phantom{b)\ 3\frac{1}{3} - \frac{1}{6}} =$ ── − ──

$\phantom{a)\ 6 - 1\frac{1}{3}} =$ ── $\phantom{b)\ 3\frac{1}{3} - \frac{1}{6}} =$ ──

$\phantom{a)\ 6 - 1\frac{1}{3}} =$ ── $\phantom{b)\ 3\frac{1}{3} - \frac{1}{6}} =$ ──

c) $5\frac{1}{3} + 2\frac{1}{4} - 1\frac{1}{2} =$ ── + ── − ──   d) $5\frac{1}{2} - 2 - \frac{7}{3} =$ ── − ── − ──

$\phantom{c)\ 5\frac{1}{3} + 2\frac{1}{4} - 1\frac{1}{2}} =$ ── + ── − ── $\phantom{d)\ 5\frac{1}{2} - 2 - \frac{7}{3}} =$ ── − ── − ──

$\phantom{c)\ 5\frac{1}{3} + 2\frac{1}{4} - 1\frac{1}{2}} =$ ── $\phantom{d)\ 5\frac{1}{2} - 2 - \frac{7}{3}} =$ ──

$\phantom{c)\ 5\frac{1}{3} + 2\frac{1}{4} - 1\frac{1}{2}} =$ ── $\phantom{d)\ 5\frac{1}{2} - 2 - \frac{7}{3}} =$ ──

4. Evaluate.

a) $3\frac{1}{2} \times 5 = \underline{} \times 5$

$= \underline{}$

$= \underline{}$

b) $8 \div 3\frac{1}{3} = \underline{} \div \underline{}$

$= \underline{} \times \underline{}$

$= \underline{}$

$= \underline{}$

5. Use the order of operations to evaluate.

a) $3\frac{1}{4} \times \left(\frac{3}{5} - \frac{2}{3}\right) \div \frac{2}{3} =$

b) $4\frac{1}{5} + \left(\frac{1}{6} + \frac{2}{3}\right) \div \frac{1}{4} =$

6. Anil ran for $\frac{1}{2}$ h on Monday, $\frac{2}{3}$ h on Wednesday, $1\frac{2}{3}$ h on Friday, and $2\frac{1}{3}$ h on Saturday. How many more hours did she run on Saturday than on Wednesday and Friday? _____

Name: _____   **Date:** _____

Chapter 3: Reviewing for the Test

3.1 Add and Subtract Fractions

page 82

1. Add or subtract.

    a) $\dfrac{1}{5} + \dfrac{1}{7} =$ ──── + ────

      $=$ ────────

      $=$ ────

    b) $\dfrac{5}{6} + \dfrac{2}{3} =$ ──── + ────

      $=$ ────────

      $=$ ────

    c) $\dfrac{7}{12} - \dfrac{7}{18} =$ ──── − ────

      $=$ ────────

      $=$ ────

    d) $\dfrac{9}{14} - \dfrac{3}{10} =$ ──── − ────

      $=$ ────────

      $=$ ────

    e) $\dfrac{5}{6} + \dfrac{3}{10} =$ ──── + ────

      $=$ ────────

      $=$ ────

    f) $\dfrac{1}{2} - \dfrac{3}{8} =$ ──── + ────

      $=$ ────────

      $=$ ────

3.2 Investigate Multiplying Fractions

page 88

2. Calculate.

    a) $\dfrac{1}{12} \times 12 =$ _____

    b) $\dfrac{5}{7} \times 7 =$ _____

    c) $\dfrac{1}{8} \times 16 =$ _____

    d) $\dfrac{1}{3} \times 12 =$ _____

    e) $\dfrac{1}{2} \times 8 =$ _____

    f) $\dfrac{1}{5} \times 25 =$ _____

3. Find each amount.

    a) $\dfrac{1}{5}$ of $10 = ──── × $____

      $= $____

    b) $\dfrac{3}{5}$ of $15 = ──── × $____

      $= $____

    c) $\dfrac{5}{9}$ of $45 = ──── × $____

      $= $____

    d) $\dfrac{3}{8}$ of $64 = ──── × $____

      $= $____

4. Jenna and her friends order a 16-slice pizza. They eat $\dfrac{3}{4}$ of the pizza.

    a) How many slices do they eat altogether? _____

    b) Her sister, Anne, eats $\dfrac{1}{2}$ of what is left. What fraction of the pizza does Anne eat? _____

Name: _____ Date: _____

3.3 Investigate Dividing Fractions

page 92

5. Divide.

a) $\dfrac{2}{3} \div \dfrac{1}{6} = \underline{} \times \underline{}$

$= \dfrac{\underline{} \times \underline{}}{\underline{} \times \underline{}}$

$= \dfrac{\underline{}}{\underline{}}$

$= \underline{}$

b) $\dfrac{4}{5} \div \dfrac{3}{10} = \underline{} \times \underline{}$

$= \dfrac{\underline{} \times \underline{}}{\underline{} \times \underline{}}$

$= \dfrac{\underline{}}{\underline{}}$

$= \underline{}$

c) $\dfrac{9}{10} \div \dfrac{2}{5} = \underline{} \times \underline{}$

$= \dfrac{\underline{} \times \underline{}}{\underline{} \times \underline{}}$

$= \dfrac{\underline{}}{\underline{}}$

$= \underline{}$

d) $\dfrac{5}{7} \div \dfrac{5}{7} = \underline{} \times \underline{}$

$= \dfrac{\underline{} \times \underline{}}{\underline{} \times \underline{}}$

$= \dfrac{\underline{}}{\underline{}}$

$= \underline{}$

6. Write each missing value.

a) $\dfrac{3}{5} \times \underline{} = \dfrac{6}{5}$ b) $\underline{} \times \dfrac{4}{7} = \dfrac{8}{7}$

c) $\dfrac{1}{4} \div \underline{} = 4$ d) $\dfrac{3}{4} \div \underline{} = \dfrac{3}{2}$

3.4 Order of Operations With Fractions

page 96

7. Evaluate.

a) $\dfrac{3}{5} \times \dfrac{5}{8} - \dfrac{1}{12} =$ b) $\dfrac{9}{12} - \dfrac{1}{3} \times \dfrac{2}{5} =$

8. Evaluate.

a) $\left(\dfrac{7}{9} + \dfrac{1}{3}\right) \times \dfrac{3}{2} = \underline{} \times \underline{}$

$= \dfrac{\underline{}}{\underline{}}$

$= \underline{}$

b) $\dfrac{5}{3} \times \left(\dfrac{5}{8} - \dfrac{3}{4}\right) = \underline{} \times \underline{}$

$= \dfrac{\underline{}}{\underline{}}$

$= \underline{}$

3.5 Operations With Mixed Numbers

page 100

9. Change each mixed number to an improper fraction.

a) $7\dfrac{5}{6} = \underline{}$ b) $5\dfrac{1}{6} = \underline{}$

10. Change each improper fraction to a mixed number.

a) $\dfrac{39}{4} = \underline{}$ b) $\dfrac{37}{3} = \underline{}$

11. Francis usually reads $8\dfrac{1}{2}$ pages of the newspaper. Today he only has time to read $2\dfrac{1}{4}$ pages. He decides to read the rest of the $8\dfrac{1}{2}$ pages tomorrow.

a) How many pages will he read tomorrow? _____

b) How many pages does he read in 6 days? _____

4.1 Explore Basic Probability

Student Text pp. 114–119

Key Ideas Review

- An **outcome** is one **possible** result of a probability experiment.

A **favourable outcome** is one that **counts** for the probability being calculated.

1. Indicate whether each is a possible outcome or a favourable outcome.

   a) A roll of a number cube gives a 1, 2, 3, 4, 5, or 6.

   b) A roll of a number cube and hoping for a 2.

- **Probability** = $\dfrac{\text{favourable outcomes}}{\text{all outcomes}}$

Hint

For this chapter, assume that the number cube has the numbers 1, 2, 3, 4, 5, and 6.

2. Select the probability of rolling a 2 using a number cube.

   a) 0   b) $\dfrac{1}{6}$ c) $\dfrac{1}{2}$ d) 1

- Probability can be described on a number line.

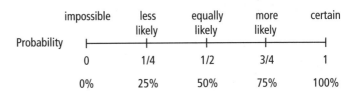

3. Indicate whether each outcome is impossible, equally likely, or certain.

   a) The sun rising in the morning.

   b) Flipping a coin and getting a head.

   c) Rolling a number cube and getting a zero.

Making Connections

You can express a probability in words, as a decimal number from 0 to 1, as a fraction from 0 to 1, or as a percent from 0% to 100%.

Example: Calculate Probability by Outcomes

There are 14 marbles in a bag: 7 clear, 3 blue, 2 black, and 2 white.

a) How many possible outcomes are there when Neko picks a marble?
b) Neko wants a black marble. How many favourable outcomes are there?
c) What is the probability of picking a black marble?
d) What is the probability of picking each colour of marble?

Name: _____   Date: _____

Solution

a) There are 14 marbles, so there are 14 possible outcomes.

b) There are two black marbles, so there are two favourable outcomes.

c) Probability(black marble) = $\dfrac{\text{favourable outcomes}}{\text{all outcomes}}$

$= \dfrac{2}{14}$

$= \dfrac{1}{7}$

d) Probability(clear marble)   Probability(blue marble)   Probability(white marble)

$= \dfrac{7}{14}$ $\qquad\qquad\qquad\qquad$ $= \dfrac{3}{14}$ $\qquad\qquad\qquad\qquad$ $= \dfrac{2}{14}$

$= \dfrac{1}{2}$ $\qquad\qquad\qquad\qquad\qquad\qquad\qquad\qquad\qquad\qquad$ $= \dfrac{1}{7}$

Practise

1. In each situation, find the total number of outcomes and the number of favourable outcomes.

   a) spinning an even number on the spinner

   b) choosing a white marble out of the bag

   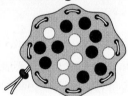

2. **a)** What is the probability of picking a white marble?

   b) What is the probability of picking a black marble?

3. Use the spinner from question 1. Mark your answers using the number line.

   ```
                    impossible   less      equally    more      certain
                                 likely    likely     likely
   Probability  ├─────────┼─────────┼─────────┼─────────┤
                    0       1/4       1/2       3/4        1
                    0%      25%       50%       75%      100%
   ```

   a) What is the probability of spinning a zero? Mark your answer using the letter A.

   b) What is the probability of spinning any number from 1 to 5? Mark your answer using the letter B.

   c) What is the probability of spinning an even number? Mark your answer using the letter C.

   d) What is the probability of spinning an odd number? Mark your answer using the letter D.

   e) What is the probability of spinning a 4? Mark your answer using the letter E.

4.2 / 4.3 Organize Outcomes and Compare Probabilities / More on Predicted Probabilities

Student Text pp. 120–130

Key Ideas Review

The key word spinner has three possible outcomes.
Determine the probability for each outcome.
Fill in the blanks with the word(s) that match the probabilities.

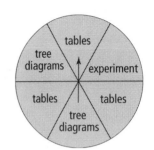

1. Results from a(n) _____ $\left(\frac{1}{6}\right)$ are used to find experimental probabilities.

2. _____ $\left(\frac{1}{3}\right)$ and _____ $\left(\frac{1}{2}\right)$ are two ways of organizing and showing the predicted probabilities.

Circle the correct word(s) in each sentence.

3. Experimental probability and predicted probability **are / are not** always the same.

4. Experimental probabilities are calculated from the results of carrying out an experiment (simulation or model). Predicted probabilities are the expected results or what you should get. The experimental probability will become **closer to / farther from** the predicted probability, the more you repeat an experiment.

Example: Use an Organizer to Show Predicted Probabilities

Kwan has four colours of pens (red, blue, purple, and green) and three colours of notepaper (yellow, green, and red). He places strips of coloured paper in a bag to match the colours for the pens and notepaper. He picks two slips of paper at a time and writes down the result. Out of 50 trials, nine pairs were the same colour.

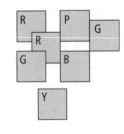

The experimental probability, based on his results, is $\frac{9}{50} = 0.18$.

a) What is the predicted probability?
b) If Kwan carried out 60 trials, how many times should he pick a pen and notepaper of the same colour?

Solution

a) Method 1: Tree Diagram

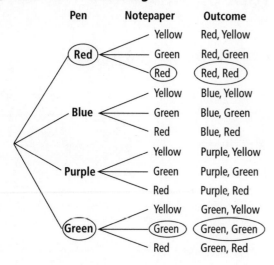

The predicted probability is $\frac{2}{12}$ or $0.1\overline{6}$.

Method 2: Table

Highlight the cells with two colours the same.

	Yellow (Y)	Green (G)	Red (R)
Red (R)	Y, R	G, R	R, R
Blue (B)	Y, B	G, B	R, B
Purple (P)	Y, P	G, P	R, P
Green (G)	Y, G	G, G	R, G

The predicted probability is $\frac{2}{12}$ or $0.1\overline{6}$.

Making Connections

The bar over the 6 in $0.1\overline{6}$ means the 6 repeats forever. You studied repeating decimals in Grade 7.

This means that Kwan should pick a pen and notepaper of the same colour two times out of 12.

b) Out of 12 picks, two should result in a pen and notepaper of the same colour: that's $\frac{1}{6}$.

Out of 60 trials, $\frac{1}{6}$ should result in a pen and notepaper of the same colour.

$\frac{1}{6} \times 60 = 10$

So Kwan should pick a pen and notepaper of the same colour 10 times out of 60.

Practise

1. Grace rolls a number cube and tosses a coin.

    a) Create a tree diagram and a table to show the possible outcomes.

    b) How many possible outcomes are there? _____

    c) What is the probability of rolling an even number and tossing tails?

    d) What is the probability of rolling an odd number and tossing heads?

2. Jon spins the spinner shown twice for a total of 32 trials. He adds the numbers for both spins. How many times should Jon get a sum of 6?

    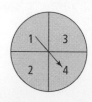

36   MHR • Chapter 4: Probability

Name: _____ Date: _____

4.4 Extension: Simulations

Student Text pp. 131–133

Key Ideas Review

- A **simulation** models a real-life situation when you cannot easily predict outcomes.
- When using a simulation, make sure the number of possible outcomes of the simulation matches the number of outcomes of the real situation.

simulation
- a probability experiment used to model a real situation

Select a possible simulation for the following situation.

Choosing one chocolate bar out of six, where there are three Big Choc, two Peanut Cupz, and one Yummies.

a) From a bag of marbles (three brown, two purple, and two yellow), choose one marble.

b) Use this spinner and make one spin.

Practise

1. Describe how you would simulate each of the following situations.

   a) There are 12 different board games. Choose one game to play. Use a spinner.

   b) A car lot has three white, four red, and seven green vehicles. Choose one car. Use a bag of marbles.

2. A restaurant gives away scratch cards printed with *W*, *I*, or *N* with every purchase. Out of every eight cards, four are *W*, one is *I*, and three are *N*.

   a) Create a spinner to simulate this promotion.

   b) How many purchases might be needed to spell *WIN*?

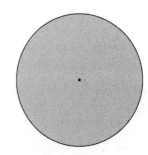

4.5 Apply Probability to Real Life

Student Text pp. 134–137

Key Ideas Review

- In some real-life games, the probability of something happening can be predicted. Predicted probability can help you make a decision in a game.
- In some real-life situations, as in sports and weather, the probability of something happening is based on experimental data.

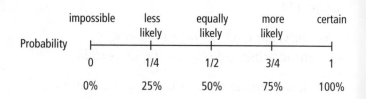

Practise

1. The weather report says that there is a 30% chance of snow.

   a) What is the probability of snow?

   b) What is the probability of no snow? Explain.

2. Cassandra noticed that her bus is on time four out of five days.

   a) What is the probability of the bus being on time today?

   b) What is the probability of the bus being late today? Explain.

3. Leslie made 14 out of 22 shots during basketball practice.

   a) What percent of the shots did she make? Round your answer to one decimal place.

   b) If she continues, what is her probability of missing the next shot? Explain.

   c) What is her probability of missing at least one of the next two shots?

Making Connections

You know that probabilities are used in predicting weather and in sports.

Read a newspaper or watch television. Where are probabilities used?

Chapter 4: Reviewing for the Test

4.1 Explore Basic Probability

page 114

1. Describe each situation as either *impossible*, *equally likely*, or *certain*.

   a) a number cube shows an even number

   b) a banana is naturally blue

   c) a baby is a girl or a boy

2. a) How many outcomes are possible when a number cube is rolled?

   b) If you are trying to roll a 5 or 6, how many favourable outcomes are there?

   c) What is the probability of rolling a 5 or 6?

3. What is the probability that

   a) you were born on a Tuesday?

   b) you were born on a weekend?

   c) you were not born on a weekend?

4.2 Organize Outcomes and Compare Probabilities
4.3 More on Predicted Probabilities

page 120

4. At a restaurant, you can have chicken, vegetables, or both on your choice of rice (white, brown, or fried).

   a) Create a tree diagram for this situation.

   b) What is the probability of having no vegetables?

5. Find the probabilities using the table.

	Coin Toss	
Spinner	H	T
A	A, H	A, T
B	B, H	B, T
C	C, H	C, T
D	D, H	D, T
E	E, H	E, T

a) What is the probability of spinning B or C?

b) What is the probability of tossing tails and spinning D?

c) What is the probability of tossing heads and spinning a consonant?

6. Syd flipped a coin 25 times. It came up tails 14 times. Which phrase describes $\frac{14}{25}$?

   A. Predicted probability of getting tails.

   B. Probability of getting tails.

   C. Experimental probability of getting tails.

4.4 Extension: Simulations

page 131

7. a) Describe a simulation for picking a letter from X, X, X, Y, Y, Y, Z, Z.

   b) What is the probability of not picking a Y?

   Probability(no Y) = _____

4.5 Apply Probability to Real Life

page 134

8. A company claims that 95 out of every 100 batteries last more than 50 hours.

   a) Determine the probability of a battery lasting more than 50 hours. Write it as a decimal and as a fraction.

   b) What is the probability of a battery lasting less than 50 hours?

5.1 Apply Ratio and Proportion

Student Text pp. 150–155

Key Ideas Review

Fill in each blank with a word, phrase, or numbers from the list.

> proportion part-to-whole part-to-part striped
> whole two 5:2 5:7

1. A _____ ratio compares _____ parts of a group.
   The ratio of _____ beads to white beads is 5:2.

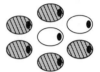

2. A _____ ratio compares a part of a group to the _____ group. The ratio of striped beads to the total number of beads is _____.

3. A _____ is a statement that two ratios are equal.   _____ = 10:4 or $\frac{5}{2} = \frac{10}{4}$

Example: Apply Ratios to Tiling

Look at the square tile pattern.

a) Write, in simplest form, the ratio that compares the number of ☐ tiles to the number of ☐ tiles.

b) Write, in simplest form, the ratio that compares the number of ▨ tiles to the total number of tiles.

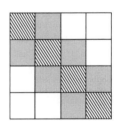

c) How many of each tile are needed to make a two-by-two repeat of the pattern?

Solution

a) There are 6 ☐ tiles and 6 ☐ tiles.

The ratio of ☐ tiles to ☐ tiles is 1:1.

> 6:6 = 1:1
> • This kind of ratio is called a proportion.

b) There are 4 ▨ tiles and 16 tiles in total.

   4:16 = 1:4

   The ratio of ▨ tiles to all tiles is 1:4.

5.1 Apply Ratio and Proportion • MHR   41

Name: _____ **Date:** _____

c) To make a two-by-two repeat of the pattern, I use four identical pattern squares.
   Multiply the number of each type of tile in the pattern by 4.

    4 × 6 tiles = 24 tiles

   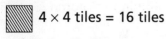 4 × 4 tiles = 16 tiles

   4 × 6 tiles = 24 tiles

   You need 24  tiles, 16 tiles, and

   24 ☐ tiles to make a two-by-two repeat of the pattern.

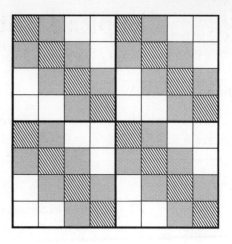

Practise

1. Look at the beads.
   Explain what each ratio represents.

   a) 1:3 _____ beads: _____ beads   b) 1:4 _____ beads: _____ beads

   c) 1:8 _____ beads: _____ beads   d) 1:2 _____ beads: _____ beads

2. Look at the bag of marbles.
   Use numbers to write each ratio in simplest form.

   a) striped marbles:white marbles ____:____

   b) striped marbles:total number of marbles ____:____

   c) grey marbles:white marbles ____:____

3. Write each ratio in simplest form.

   a) 6:12 = ___:___   b) 21:49 = ___:___   c) 24:96 = ___:___   d) 18:54 = ___:___

4. Write the missing term in each proportion.

   a) $\frac{}{5} = \frac{6}{15}$ b) $\frac{4}{} = \frac{20}{25}$ c) $\frac{3}{7} = \frac{}{49}$ d) $\frac{6}{6} = \frac{18}{}$

5. Look at the shape pattern. It has 9 blocks in total.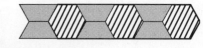

   a) Write, in simplest form, the ratio that compares the number of striped blocks with the number of shaded blocks.

   _____:_____

   b) Write, in simplest form, the ratio that compares the number of striped blocks with the total number of blocks.

   _____:_____

5.2 Explore Rates

Student Text pp. 156–161

Key Ideas Review

Draw a line to match each item in Column A with its description in Column B.
Then draw a line to match each description in Column B with its example in Column C.

A	B	C
1. A rate	a) A rate in which the second term is 1	i. 50¢ per orange
2. A unit rate	b) A unit rate involving prices	ii. $2.50 for 3 L
3. A unit price	c) A comparison of quantities measured in different units	iii. 10 km/h

Example 1: Applying Rates to Earnings

On Saturday, James earns $25 for cutting his neighbour's lawn in 5 h.

a) Write a unit rate that describes how much James is paid.
b) How much will James earn if he spends 7 h cutting lawns?

Strategies
Make a picture or diagram.

Solution

a) James earned $25 in 5 h. Write this rate in fraction form.
$$\frac{\$25}{5\ h} = \frac{\$5}{1\ h}$$

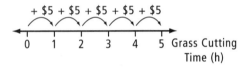

As a unit rate, James is paid $5/h.

b) James earns $5/h. To find how much he will earn in 7 h, multiply by 7. James will earn $35 for 7 h of cutting lawns.

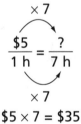

Example 2: Applying Rates to Shopping

Which jar of jam is the better buy? Use unit rates to compare.

Solution

Tasty Jam:

Unit price = $\frac{\$3.25}{250\ g}$
       = $0.013/g

Jane's Jam:

Unit price = $\frac{\$2.99}{175\ g}$
       = $0.017\ 09/g

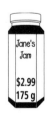

Tasty Jam costs $0.013/g or 1.3¢/g.   Jane's Jam costs $0.017 09/g or 1.7¢/g.

The unit price for Tasty Jam is less than the unit price for Jane's Jam.
Tasty Jam is the better buy.

Name: _____   Date: _____

Practise

1. Katrina earns $9.00/h working in a clothing store. How much will she earn in

   a) a 5-h shift? $_____   b) a 40-h work week? $_____

2. Jinji is paid $20.25/h. How much does he earn in

   a) a 7-h day? $_____   b) a 37-h work week? $_____

3. Find the unit rate in each situation. Answer to two decimal places.

   a) Elise roller skated 3 km in 30 min. _____

   b) Jojo earned $42 in 4 h. $_____

   c) Lila paid $230 to rent a car for 15 h. $_____

4. Find the unit price of each item. Answer to two decimal places.

   a)    b)    c)

   _____¢/g   _____¢/bar of soap   _____¢/mL

5. How much does each item cost?    $0.36/kg    $0.30/100 ml

   a) 2-kg bag of carrots $_____

   b) a 1-L bottle of water $_____

6. Bill types 150 words in 5 min. Errol types 200 words in 8 min. Fill in the blanks.

   a) Bill types _____ words/min and Errol types _____ words/min.
   Therefore, _____ types faster than _____ types.

   b) How many minutes will it take each person to type 2000 words? Answer to one decimal place.
   Bill: _____ min   Errol: _____ min

5.3 Apply Percents to Sales Taxes and Discounts

Student Text pp. 162–165

Key Ideas Review

Mark each statement **T** (for true) or **F** (for false). If the statement is false, cross out one word and write another word above it to make the statement true.

> **Making Connections**
>
> You could do some research by visiting a clothing or electronics store and asking questions.

____ 1. Retailers don't have to collect taxes.

____ 2. In Ontario, three taxes are added on to purchases.

____ 3. A discount is an amount subtracted from the regular price of an item to give a sale price.

____ 4. To calculate the final price, subtract any discounts after adding the sales taxes.

Example: Calculate PST and GST

The regular price for a skateboard is $89.50. It is on sale at a 30% discount.

a) What is the sale price of the skateboard?
b) How much will the skateboard cost when all taxes are included?

> **Hint**
>
> In Ontario, the GST is 7% or 0.07. The PST is 8% or 0.08.

Solution

a) **Method 1: Find the Discount First, Then Subtract**
Discount = 30% of $89.50
= 0.30 × $89.50
= $26.85
Sale price = regular price − discount
= $89.50 − $26.85
= $62.65

Method 2: Find the Sale Price Directly
Sale price = regular price − discount
= 100% − 30%
= 70%
Sale price = 70% of $89.50
= 0.70 × $89.50
= $62.65

Either way, the sale price of the skateboard is $62.50.

b) **Method 1: Find Each Tax Separately**
Provincial Sales Tax (PST) is 8% of $62.65.
PST = 0.08 × $62.65
= $5.01
Goods and Services Tax (GST) is 7% of $62.65.
GST = 0.07 × $62.65
= $4.39
Add to find the total cost.
Skateboard $62.65
PST $5.01
GST $4.39
TOTAL $72.05

Method 2: Find Both Taxes Together
Adding 8% of a number and 7% of a number is the same as adding 15% of a number.
Total tax = 15% of $62.65
= 0.15 × $62.65
= $9.40
Add to find the total cost.
Skateboard $62.65
PST & GST $9.40
TOTAL $72.05

Either way, the cost of the skateboard is $72.05.

Name: _____ Date: _____

Practise

1. Find the 7% GST on each item.

a)

b)

c)

2. Find the 8% PST on each item.

a)

b)

c)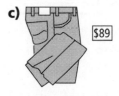

3. Fill in the PST and GST on each item.

Item	DVD	Movie Ticket	Scarf
Price	$23.85	$12.00	$15.40
PST (8%)	$	$	$
GST (7%)	$	$	$

4. Find the total cost of each item, including the PST and GST.

Item	Roll of Wallpaper	Bunch of Flowers	Car
Price	$16.00	$3.99	$25 000
PST (8%)	$	$	$
GST (7%)	$	$	$
Total Cost	$	$	$

5. A sports store is having a sale. All items are discounted by 30%. Calculate the sale price of each item. Then calculate the total price of each item, including PST and GST.

Item	Baseball Bat	Basketball	Pair of Skis
Regular Price	$23.44	$34.99	$230.00
Sale Price	$	$	$
PST (8%)	$	$	$
GST (7%)	$	$	$
Total Cost	$	$	$

5.4 Apply Percent to Commission

Student Text pp. 160–169

Key Ideas Review

Check each job that would likely pay commission.

____ selling cars   ____ designing cars

____ computer programming   ____ selling computers

____ receptionist   ____ selling magazine subscriptions

Example: Calculate Commission Earnings and Apply Rates to Commission

Kai has a summer job selling magazine subscriptions door-to-door. He is paid a commission of 35% of sales. Each new subscription sells for $30.

a) One day, he sells 12 subscriptions. What is Kai's commission for that day?
b) Kai hopes to earn $1400 selling magazines. What total value must he sell?

Solution

a) First, calculate the total sales.
Total sales = 12 × $30
            = $360
Then, calculate his commission.
Commission = 35% × $360
           = 0.35 × $360
           = $126
Kai's commission for that day is $126.

Estimate: 35% is a bit more than one-third (33%). One-third of $360 is $120.

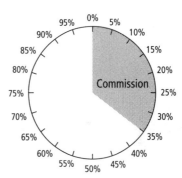

b) 35% commission means that Kai earns $35 for every $100 of sales.

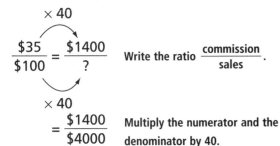

Write the ratio $\frac{\text{commission}}{\text{sales}}$.

Multiply the numerator and the denominator by 40.

Kai must sell $4000 worth of magazines.

Total Sales

Making Connections

Research some jobs that pay a salary and some that pay a commission.

Would you prefer being paid a salary or a commission? Give reasons for your choice.

Practise

1. Jasmine earns a 15% commission selling hats in a booth. How much does she earn for each day's sales?

   a) Friday: sales, $200; commission, $_____

   b) Saturday: sales, $260; commission, $_____

   c) Sunday: sales, $50; commission, $_____

2. A textbook salesperson earns a 10% commission on sales. Estimate, then calculate, the commission earned on each month's order.

   a) May: sales, $20 050; commission estimate, $_____; calculation, $_____

   b) June: sales, $15 750; commission estimate, $_____; calculation, $_____

   c) July: sales, $4100; commission estimate, $_____; calculation, $_____

3. Lana works in a clothing store. Her commission is 22% of sales. One morning, she sold these items. What were her earnings for that morning?

   $_____

$225 $360 $70

4. Alannah earns a 25% commission on computer sales. Fill in her sales for each week.

Week	Sales	Earnings
1	$	$1500
2	$	$250
3	$	$800

5. Brooke earns a 30% commission on furniture sales. Complete his sales record for the week.

Day	Sales	Earnings
Tuesday	$450.00	$
Wednesday	$	$405.00
Thursday	$160.00	$
Friday	$	$45.00
Saturday	$2300.00	$
Total	$	$

Name: _____   Date: _____

5.5 Calculate Simple Interest
Student Text pp. 170–173

Key Ideas Review

The formula for simple interest is $I = P \times r \times t$.

Enter the correct letter in each space.

1. _____ is the principal in dollars.
2. _____ is the time in years.
3. _____ is the interest rate per year in decimal form.
4. _____ is the interest in dollars.

Example 1: Calculate Simple Interest

Johdi buys a $500 Canada Savings Bond that pays $7\frac{1}{4}\%$ per year simple interest. Johdi can cash the bond in 10 years when it matures.

a) How much interest will the bond earn after 10 years?
b) What will the total value of the bond be after 10 years?

Solution

a) Use the simple interest formula $I = P \times r \times t$.
   $I = P \times r \times t$
   $I = \$500 \times 0.0725 \times 10$
   $I = \$362.50$ Use two decimal places for money.
   The bond will earn $362.50 after 10 years.

   > I is the interest in dollars. This is the amount you want to find.
   > $P = \$500$
   > r is in decimal form.
   > In decimal form, the interest rate per year is
   > $r = 7\frac{1}{4}\% = 0.0725$.
   > $t = 10$ years

b) Total amount = principal + interest
   $A = \$500 + \$362.50$
   $A = \$862.50$
   The total value of the bond after 10 years is $862.50.

Example 2: Calculate Simple Interest for Less Than One Year

Willy borrows $300 and agrees to pay 6% interest per year. If Willy repays the loan after 3 months, how much does he owe?

Solution

Step 1: Find the interest.

The interest rate is 6% per year or 0.06. Time needs to be in years. 3 months is $\frac{3}{12} = \frac{1}{4}$ year. Use $t = 0.25$.

$I = P \times r \times t$
$I = \$300 \times 0.06 \times 0.25$
$I = \$4.50$

Name: _____   Date: _____

Step 2: Add the principal and interest to find the total amount owing.

Amount = principal + interest
   A = $300.00 + $4.50
   A = $304.50

Willy will owe $304.50.

Practise

1. You are going to invest $400 in a savings account. Fill in the interest you would earn in each case.

   a) 1 year at 2.5% interest $_____

   b) 2 years at 3.25% interest $_____

   c) 3 years at 2.5% interest $_____

2. You are going to invest $2000 in a money market account. Fill in the interest you would earn in each case. Fill in the total amount of money you would have.

Years in Account	Interest Rate	Interest Earned	Total Amount
1	2.5%	$	$
2	3.25%	$	$
3	2.25%	$	$

3. Fill in the total interest that each bond would pay. Fill in the total value of each bond at maturity.

Cost of Bond	Interest Rate	Years to Maturity	Interest Earned	Total Value
$100	6.5%	4	$	$
$250	8.25%	10	$	$
$500	10.5%	6	$	$
$1000	7.65%	3	$	$

4. Jerry deposits $430 in an account that earns 3.25% interest per year.

   a) What is the interest after 3 months? $_____

   b) What is the value of the deposit after 3 months? $_____

5. Cleo deposits $870 in an account that earns 5.20% interest per year.

   a) What is the interest after 9 months? $_____

   b) What is the value of the deposit after 9 months? $_____

6. Jack borrows $500 at 8.5% interest per year. He plans to repay the loan and interest after 9 months. What will be the total amount that he has to pay back? $_____

Name: _____   Date: _____

Chapter 5: Reviewing for the Test

5.1 Apply Ratio and Proportion

page 150

1. Write each ratio in simplest form.

    a) 8:12 = ___:___   b) 28:49 = ___:___

    c) 32:96 = ___:___   d) 17:51 = ___:___

2. Write the missing term in each proportion.

    a) $\dfrac{}{5} = \dfrac{10}{25}$ b) $\dfrac{8}{} = \dfrac{24}{30}$

3. The shape pattern has 12 blocks in total.

    Write the ratio, in simplest form, that compares the number of

    a) striped blocks with shaded blocks.

    _____:_____

    b) striped blocks with the total number of blocks.

    _____:_____

5.2 Explore Rates

page 156

4. Katrina works in a clothing store for $11.50/h. How much will she earn in

    a) 5 h? $_____

    b) 40 h? $_____

5. Find the unit rate in each case.

    a) Jill biked 10 km in 40 min. _____

    b) Megan earned $60 in 4 h. $_____

    c) Mark paid $24 to rent a canoe for 3 h.
    $_____

5.3 Apply Percents to Sales Taxes and Discounts

page 162

6. Find the 7% GST on each item.

    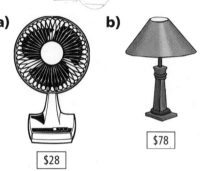

    a) $28   b) $78

    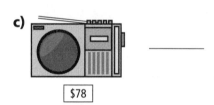

    c) $78

7. Find the 8% PST on each item in question 6.

    a) _____   b) _____

    c) _____

8. Fill in the PST and GST on each item.

Item	CD	Tent
Price	$16.85	$126.00
PST	$	$
GST	$	$

9. Find the total cost of each item, including the PST and GST.

Item	Can of Paint	Cat Toy
Price	$27.34	$6.95
PST	$	$
GST	$	$
Total Cost	$	$

5.4 Apply Percent to Commission

page 166

10. Tony earns a 15% commission selling popcorn in a booth. How much does he earn for each day's sales?

a) Tuesday: sales, $120; commission, $_____

b) Wednesday: sales, $73; commission, $_____

c) Thursday: sales, $80; commission, $_____

11. A sales representative earns a 10% commission on sales. Estimate, then calculate, the commission earned on each month's order.

a) January: sales, $10 500;
Commission estimate, $_____;
Calculation, $_____

b) February: sales, $5750;
Commission estimate, $_____;
Calculation, $_____

12. Chenisse sells tires. Her commission is 22% of sales. One afternoon, she sold two sets of four tires for $300 per set, and a single tire for $80. What were her earnings for that afternoon?

$_____

13. Rupert earns a 20% commission on computer sales. Fill in his sales for each week.

Week	Sales	Earnings
1	$	$1800
2	$	$200
3	$	$1000

5.5 Calculate Simple Interest

page 170

14. You are going to invest $2500 in a savings account. Complete the table.

Years in Account	Interest Rate	Interest Earned	Total Amount
1	2.5%	$	$
2	3.25%	$	$
3	2.25%	$	$

15. Complete the table for these bonds.

Cost of Bond	Interest Rate	Years to Maturity	Interest Earned	Total Value
$100	7.5%	4	$	$
$200	6.25%	10	$	$
$500	9.5%	6	$	$
$1000	7.60%	2	$	$

16. Ram deposits $530 in an account that earns 3.5% interest per year.

a) What is the interest after 3 months?
$_____

b) What is the value of the deposit after 3 months? $_____

Study Skills

There are several methods to solve a problem involving discounts. Write a question involving percent discounts. Solve it using one method and check using another. Exchange questions with a study partner. Have your partner verify your work by solving the problem using the method you used to check your work.

6.1 Identify Patterns

Student Text pp. 182–185

Key Ideas Review

- Some patterns are based on geometric shapes or lines.
- Other patterns are based on number operations.
- To identify a pattern:
  - Find the first shape or number.
  - Describe how new shapes or numbers are generated.
  - Look for repeated sets of operations.

Examine this pattern of numbers.

35, 31, 27, 23, ...

1. Select the statement that describes the number pattern.

   A. The next number is 4 more than the one before.

   B. The next number is 4 times the one before.

   C. The next number is 4 less than the one before.

2. Select the next three numbers in the sequence.

   A. 27, 31, 35   **B.** 132, 528, 2112   **C.** 19, 15, 11

Example 1: Extend a Fractal Tree

Fractal trees show a visual pattern.
Examine this fractal tree.

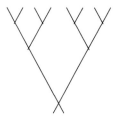

fractal
- a pattern of shapes, lines, or colours that gets smaller as it repeats

a) Describe how new branches are created.
b) Extend the branches to one more stage.

Solution

a)

b)

Begin with a "V."
For each stage, add to the end of each branch
a "V" that is half as long as those in the previous stage.

Example 2: Identify a Number Pattern

Describe each number pattern and write the next three terms.

a) 4, 7, 10, 13, …   b) 1, 3, 9, 27, …   c) 30, 27, 24, 21, …

Solution

a) Start with 4. Add 3 repeatedly.
The next three terms are 16, 19, and 22.

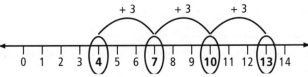

b) Start with 1. Repeatedly multiply by 3.

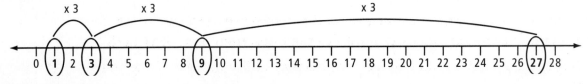

The next three terms are 81, 243, and 729.

Practise

1. Describe each number pattern. Then write the next three numbers.

   a) 22, 19, 16, 13, …   b) 4, 8, 16, 32, …

   c) 0.5, 0.9, 1.3, 1.7, …   d) 33, 30, 27, 24, …

2. Describe how new branches are created. Extend the branches to one more stage.

3. The Sierpinski gasket is made using the steps shown. Explain how to extend this fractal pattern.

 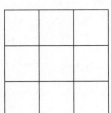 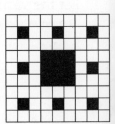

6.2 Define Patterns Using Algebra

Student Text pp. 186–191

Key Ideas Review

- Patterns can be modelled using equations.
- To develop a formula:
  - Create a table to identify a pattern.
  - Write a number sentence algebraically.
- You can use an expression for the value of the nth term to make predictions about a pattern.

Example: Apply Number Patterns

The Fishy Fish House is having a sale: With the purchase of a $5 can of fish food, a customer can buy fancy goldfish for the discounted price of $4 each.

a) Describe the relationship between the number of fish purchased and the total purchase price.
b) Use a table of values to determine the pattern between the total cost and the number of goldfish.
c) Write an equation to model the total cost. Define your variables.
d) Find the total cost of purchasing seven goldfish at the sale.

variable
- a letter or symbol that represents a number or numbers

Solution

a) The total cost will be $5 for the fish food plus $4 times the number of fish.

b)
Number of Fish	Total Cost ($)
1	1(4) + 5 = 9
2	2(4) + 5 = 13
3	3(4) + 5 = 17
4	4(4) + 5 = 21

Literacy Connections

In this pattern, 9 is the first number or term of the pattern, 13 is the second term, and so on with n as the nth term.

c) Let the variable n represent the number of fish purchased.
Let the variable T represent the total cost in dollars.
The equation is $T = 4n + 5$.
Total cost / Cost of one fish / Number of fish purchased / Cost of fish food

d) $T = 4n + 5$
$T = 4(7) + 5$
$T = 33$
The total cost of purchasing seven fish and the food is $33.

Practise

1. Describe each number pattern. Then determine an expression for the *n*th term. Write the next three terms.

   a) 4000, 2000, 1000, 500, ...

   b) $\dfrac{1}{3}, \dfrac{1}{4}, \dfrac{1}{5}, \dfrac{1}{6}, \ldots$

   c) $15\dfrac{1}{4}, 13\dfrac{1}{4}, 11\dfrac{1}{4}, 9\dfrac{1}{4}, \ldots$

   d) $2\dfrac{2}{7}, 3\dfrac{4}{7}, 4\dfrac{6}{7}, 6\dfrac{1}{7}, \ldots$

Apply

2. Karla can purchase school T-shirts for $9.50 each.

   a) Describe the relationship between the total cost and the number of T-shirts purchased.

   $9.50

   b) Complete the table.

Number of T-shirts	1	2	3	4
Total Purchase Cost ($)	9.50			

   c) Use the variables *n* for the number of T-shirts purchased and *C* for the total purchase cost. Write an equation to model the total cost of *n* T-shirts.

   d) Determine the cost of 13 T-shirts.

3. Perfect Pizza charges $7.00 for a medium cheese pizza and 75¢ for each additional topping.

   a) Write a formula for the cost of a medium pizza with *n* toppings. Define your variables.

   b) Determine the total cost of a medium pizza with six additional toppings.

56 MHR • Chapter 6: Patterning and Algebra

Name: _____ Date: _____

6.3 Explore Relationships on a Grid

Student Text pp. 192–197

Key Ideas Review

Fill in the table with an item from the list. Some choices may not be used.

5	4	3	(4, 9)	(7, 3)	(7, 9)	(3, 7)
1	9	7	2	(4, 3)	(9, 4)	(3, 9)

1. Plot these ordered pairs on the grid.

   a) (0, 1)   b) (1, 3)

2. Fill in the table of values for the graph.

x	0	1	2	3	4
y					

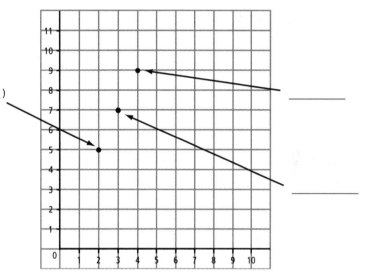

Practise

1. Kaycee designs greeting cards. She charges a flat fee of $15 plus $1.25 per card.

   a) Complete the tables of values.

   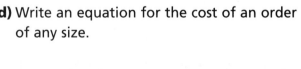

Number of Cards	10	15	20	25	30	35	40
Cost ($)							

   b) Write the ordered pairs. Plots the points on the grid. Label your graph carefully.

   c) Describe the relationship between the cost and the number of cards purchased.

   d) Write an equation for the cost of an order of any size.

   e) Determine the cost of an order of 50 cards.

6.4 Apply Patterning Strategies

Student Text pp. 198–203

Practise

1. You are planning the obstacle course event for the school fun day. After each pair of participants races, the winner moves on to the next round.

   Strategies
   Work backward. Start with the winner and use a tree diagram to show each round.

   a) How many participants can there be if the event needs five rounds to declare a winner?

   b) How many participants can there be if it takes n rounds to determine the winner?

2. A pattern of triangles can be made using toothpicks.

   a) Determine an equation for the perimeter, P, of the nth triangle.

   b) What is the perimeter of the sixth triangle?

3. Tables are being set up in the gymnasium for the graduation party. As shown in the diagram, the long sides of each table seat 5 people in total, and the short sides each seat 1 person.

   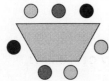

   a) Write a formula for the number of people, P, that can be seated at n tables arranged in one straight line.

   b) The longest arrangement that will fit in the square gym is 7 tables long. Draw a seating plan for all 111 people expected to attend.

Chapter 6: Reviewing for the Test

6.1 Identify Patterns

page 182

1. Describe each number pattern. What are the next three numbers?

   a) 3, 9, 15, 21, …

   b) 76, 70, 64, 58, …

   c) 0.1, 0.5, 2.5, 12.5, …

   d) 13.8, 12.1, 10.4, 8.7, …

2. a) Describe how new branches are created.

   b) Extend the branches to one more stage.

6.2 Define Patterns Using Algebra

page 186

3. For each sequence, determine an expression for the nth term.

   a) $4 \times 2, 4 \times 2^2, 4 \times 2^3, …$

   b) 8, 14, 20, 26, …

4. Examine the pattern of cubes with side lengths of 1 unit.

   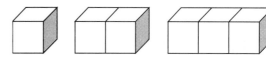

   a) Describe the relationship between the number of cubes and the surface area.

   b) Write an equation for the surface area, A, of the row of n cubes.

   c) Use your equation to find the surface area of the row of six cubes.

6.3 Explore Relationships on a Grid

page 192

5. a) Complete the table of values for the equation $y = 10x + 1$.

x	0	1	2	3	4
y					

b) List the ordered pairs.

c) Plot your ordered pairs on the grid. Label your graph carefully.

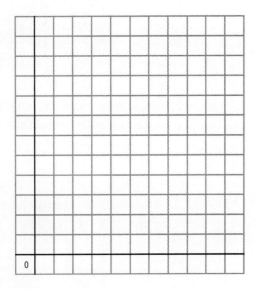

d) Describe the pattern.

6.4 Apply Patterning Strategies

page 198

6. A pattern is made from 1-cm square tiles.

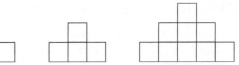

a) By how much does the perimeter increase each time a new row is added?

b) Describe the perimeter, P, of a tile pattern made with n rows.

c) Find the perimeter of a pattern made with 10 rows of tiles.

7. In the class spell-off, pairs of students compete against each other. The winning student from each pair goes on to the next round. How many rounds are needed to declare a winner in a class of 32?

7.1 Pattern With Powers and Exponents

Student Text pp. 214–219

Key Ideas Review

Fill in each blank with a word from the list. Each choice will be used more than once.

> base   exponent   power

1. Label the diagram.

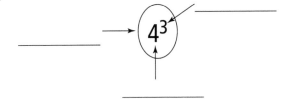

2. A _____ expresses repeated multiplication.

3. The _____ is the factor being multiplied.

4. The _____ tells how many factors you multiply.

5. A _____ is a number in exponential form.

Literacy Connections

Reading Powers
Powers can be described in terms of the base and exponent.

The power 4^5 can be read as
- four to the fifth
- four to the exponent five
- the fifth power of four

Example 1: Write Powers

Express as a power.

a) $14 \times 14 \times 14$
b) $6.1 \times 6.1 \times 6.1 \times 6.1 \times 6.1 \times 6.1$

Solution

a) $14 \times 14 \times 14 = 14^3$

b) $6.1 \times 6.1 \times 6.1 \times 6.1 \times 6.1 \times 6.1 = 6.1^6$

Example 2: Write Powers

Write 2401 as a power of 7.

Solution

$2401 = 7 \times 343$
$ = 7 \times 7 \times 49$
$ = 7 \times 7 \times 7 \times 7$
$ = 7^4$

Name: _____ Date: _____

Example 3: Evaluate Powers

Evaluate.

a) 6^4 b) 4.9^3

Solution

a) $6^4 = 6 \times 6 \times 6 \times 6$
$= 36 \times 36$
$= 1296$

b) Use a calculator.
[C] 4.9 [y^x] 3 [=] 117.649

Practise

1. Express as a power. Do not evaluate.

   a) $13 \times 13 \times 13 \times 13$ b) $5 \times 5 \times 5$

   c) $7 \times 7 \times 7 \times 7 \times 7 \times 7$ d) $5.6 \times 5.6 \times 5.6 \times 5.6 \times 5.6$

2. Write each power.

   a) 243 as a power of 3   b) 625 as a power of 5

   c) 512 as a power of 2   d) 512 as a power of 8

3. Evaluate.

   a) 4^4 b) 9^5

   c) 5.1^4 d) 0.6^3

Apply

4. Find the volume of a cube with each edge length.

   a) 3 cm   b) 4.2 mm

Hint

Volume of cube
= length × width × height

62 MHR • Chapter 7: Exponents

7.2 Order of Operations With Exponents

Student Text pp. 220–225

Key Ideas Review

Fill in each blank with a word from the list.

> meanings   brackets   fraction   division   grouping

1. A __ __ __ __ __ __ __ __ __ bar has two __ __ __ __ __ __ __ __ __. It acts as a __ __ __ __ __ __ __ __ __ symbol and also as a __ __ __ __ __ __ __ __ __ symbol, like __ __ __ __ __ __ __ __.

2. Place the letters from the boxes in the correct order of operations. Write the word each letter stands for on the line beside it.

Example 1: Evaluate Expressions with Exponents

Evaluate $3^3 + 4(6^2 + 1) \div 2$. Justify each step.

Solution

$3^3 + 4(6^2 + 1) \div 2$	Brackets: exponent.
$= 3^3 + 4(36 + 1) \div 2$	Brackets: add.
$= 27 + 4(37) \div 2$	Multiply.
$= 27 + 148 \div 2$	Divide.
$= 27 + 74$	Add.
$= 101$	

Example 2: Evaluate Expressions with Fraction Bars

Evaluate $\dfrac{14 + 8}{11} + 3(2)$. Justify each step.

Name: _____ Date: _____

Solution

Method 1: Keep the Fraction Bar

$\dfrac{14+8}{11} + 3(2)$

$= \dfrac{(14+8)}{11} + 3(2)$ Brackets.

$= \dfrac{22}{1} + 3(2)$ Divide. Multiply.

$= 2 + 6$ Add.

$= 8$

Method 2: Eliminate the Fraction Bar

$\dfrac{14+8}{11} + 3(2)$

$= (14+8) \div 11 + 3(2)$ Brackets.

$= 22 \div 11 + 3(2)$ Divide. Multiply.

$= 2 + 6$ Add.

$= 8$

Practise

1. Evaluate. Justify each step.

a) $2^5 + 4(3 + 2) - 6$

b) $2^4 \div (2^2 + 5 - 1)$

c) $6(7 + 2)^2$

d) $7 \times 5 - (3^5 - 2^2)$

2. Evaluate. Justify each step.

a) $16 - \dfrac{11 + 3}{7}$

b) $\dfrac{4^2 + 2}{10 - 1} + 8$

Apply

3. a) Write an expression that represents the area of this shape.

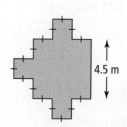

4.5 m

b) Calculate the area.

64 MHR • Chapter 7: Exponents

7.3 Discover Scientific Notation

Student Text pp. 226–231

Key Ideas Review

Draw a line to show where each study note belongs.

  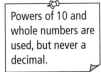

Always a product of two numbers. | Shows the value of each digit in a number. | Powers of 10 and whole numbers are used, but never a decimal. | Useful for writing large numbers. | Uses powers of 10 and numbers no smaller than 1 and less than 10.

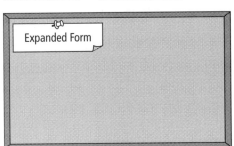

Expanded Form

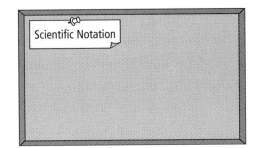

Scientific Notation

Example 1: Write Numbers in Expanded Form Using Powers of 10

Use powers of 10 to write 6982 in its expanded form.

Solution

$6982 = 6000 + 900 + 80 + 2$
$= 6 \times 10^3 + 9 \times 10^2 + 8 \times 10^1 + 2$

Example 2: Write Numbers in Scientific Notation

Write 821 000 in scientific notation.

Solution

Method 1: Write as a Product
$821\ 000 = 821 \times 1000$
$= 8.21 \times 100 \times 1000$
$= 8.21 \times 10^5$

Method 2: Move the Decimal Point
The decimal point moves 5 places, so
$821\ 000 = 8.21 \times 10^5$

Example 3: Convert to Standard Form

Convert 9.49×10^7 to standard form.

Solution

$9.49 \times 10^7 = 9.49 \times 10\ 000\ 000$
$= 94\ 900\ 000$

Name: _____ Date: _____

Practise

1. Use powers of 10 to write each number in expanded from.

   a) 468

   b) 50 069

   c) 92 931

   d) 727 543

2. Write each number in scientific notation.

   a) 49 500 000

   b) 23 100

   c) 600 000 000

   d) 7 410 000

   Literacy Connections

   You can read 1.61×10^4 as
   - one decimal six one times ten to the fourth
   - one decimal six one times ten to the exponent four
   - one decimal six one times the fourth power of ten

3. Express each number in standard form.

   a) 1.61×10^4

   b) 7.239×10^6

   c) 4.5×10^8

   d) 5×10^4

Apply

4. Without using a calculator, arrange the numbers in order from least to greatest.

   a) 6.21×10^4

   $6 \times 10^4 + 2 \times 10^3 + 1$

   6021

   Making Connections

   When you look at population data for towns, often the tables will have a heading like "Population (100 000s)." Why is the expanded form of a number used?

   If the population of Kingston, Ontario, is listed in the data table as 11.3, how many people live in Kingston?

   b) $1 \times 10^4 + 7 \times 10^3 + 7 \times 10^2$

   1.77×10^3

   1777

Name: _____ Date: _____

7.4 Solve Fermi Problems

Student Text pp. 232–235

Key Ideas Review

Circle the word or phrase that will complete each sentence correctly.

1. The answer to a Fermi problem is an **exact solution** / **estimate**.

2. A Fermi problem may have **many solutions** / **only one solution**.

3. To solve a Fermi problem, you **research missing information** / **use only the information given** and make assumptions.

Practise

1. Estimate the number of minutes it would take you to walk across Canada. Complete the solution.

Understand
- Estimate the distance across Canada.

Plan
- Estimate how long it would take you to walk 1 km.
- Multiply the distance across Canada by the time it would take you to walk 1 km.

Do It!
- Assume the distance across Canada is from Victoria, British Columbia, to St. John's, Newfoundland.

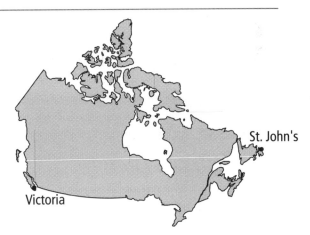

Strategies
Find needed information in an atlas or on the Internet.

Select a reasonable estimate for the distance across Canada.
a) 7314 km
b) 7314 m
c) 7314 cm

Strategies
Make an assumption.

Estimate how long it would take you to walk 1 km.
a) 1 h
b) 10 min
c) 10 s

7.4 Solve Fermi Problems • MHR **67**

Name: _____   Date: _____

Strategies
Make an assumption.

Select the best estimate of the time it would take you to walk across Canada.

a) 731.4 h
b) 12.9 h
c) 1219 h

Look Back

- Try solving the problem using another strategy.

2. Explain and justify the process you used to get each result. How many hockey sticks would you need to be able to make a solid line around the equator?

Making Connections

Why is scientific notation useful in solving Fermi problems?

Chapter 7: Reviewing for the Test

7.1 Pattern With Powers and Exponents

page 214

1. Express as a power.

   a) $7.5 \times 7.5 \times 7.5$

   b) $3 \times 3 \times 3 \times 3 \times 3 \times 3 \times 3 \times 3$

2. Write each power.

   a) 343 as a power of 7

   b) 128 as a power of 2

   c) 2744 as a power of 14

3. Evaluate.

   a) 11^3

   b) 7^4

   c) 3.1^6

   d) 0.9^3

7.2 Order of Operations With Exponents

page 220

4. Evaluate. Justify each step.

   a) $(10 - 6)^3 - 16 \div 4$

   b) $4 \times 3 + (7 - 5)^2$

5. Evaluate. Justify each step.

   a) $15 + \dfrac{9^2 \div 3}{2 + 1}$

   b) $\dfrac{3^3 + 6}{2^4 - 5}$

Name: _____ Date: _____

7.3 Discover Scientific Notation

page 216

6. Use powers of 10 to write each number in expanded form.

   a) 964

   b) 6731

   c) 8672

7. Write each number in scientific notation.

   a) 335 000

   b) 4 000 000

   c) 611 000 000 000 000

8. Express each number in standard form.

   a) 9.2×10^7

   b) 4.22×10^3

9. Without using a calculator, arrange the numbers in order from least to greatest.

   2987

   2.9×10^3

   $2 \times 10^3 + 9 \times 10^2 + 8 \times 10^1$

7.4 Solve Fermi Problems

page 232

10. Describe how you would solve the following problems. In your description, list the missing information you would need and the assumptions you would make.

    a) How many of your schoolyards would cover your province?

    b) How many times will you answer the phone in your lifetime?

8.1 Recognize and Sketch Three-Dimensional Figures

Student Text pp. 244–247

Key Ideas Review

A three-dimensional object can be represented by the side, top, and front views. Label the views of this three-dimensional object.

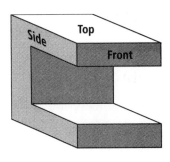

A. _____   B. _____   C. _____

Practise

1. Sketch the front, side, and top views of this three-dimensional figure.

   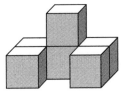

2. Identify the geometric figure from the front, top, and side views.

   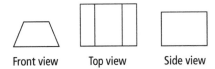

   Front view   Top view   Side view

3. The top view of a pyramid is shown. What kind of pyramid is the figure?

   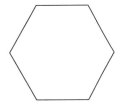

8.1 recognize and Sketch Three-Dimensional Figures • MHR 71

8.2 Build Models of Three-Dimensional Figures

Student Text pp. 248–252

Key Ideas Review

Move from letter to letter along the edges to spell the missing words. You may use a letter more than once. The first letters of each word are indicated.

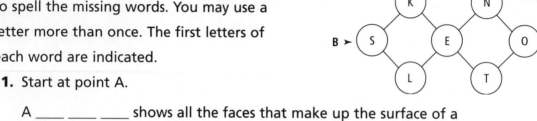

1. Start at point A.

   A ___ ___ ___ shows all the faces that make up the surface of a three-dimensional figure.

2. Start at point B.

   A ___ ___ ___ ___ ___ ___ ___ ___ is a frame formed by joining the edges of a three-dimensional figure.

Example: Nets and Skeletons

Sketch the net and skeleton of a pentagonal prism. How many faces, edges, and vertices does it have?

Solution

Net and Faces
A pentagon has five sides, so the prism will have five rectangular faces.
Prisms have two bases, so there are two pentagonal faces.

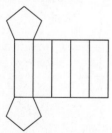

Count the faces.
There are seven faces.

Skeleton, Edges, and Vertices
Start with two pentagons.

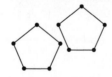

Form the edges between the bases.

Count the edges and vertices.
There are 15 edges and 10 vertices.

Name: _____ Date: _____

Practise

1. Which of the following is a net for a square prism? Why aren't the other two?

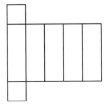 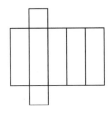 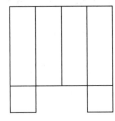

A.   B.   C.

2. Sketch the skeleton of each three-dimensional figure. How many edges and vertices does each figure have?

   a) square-based pyramid   b) parallelogram-based prism

   c) hexagonal pyramid   d) hexagonal prism

3. a) How many edges and vertices does a pentagonal prism have?

   b) How is the number of sides of a pyramid's base related to the number of edges and vertices?

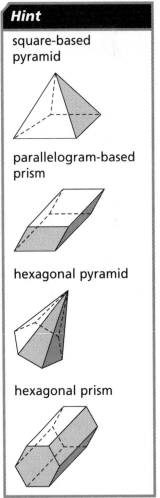

Hint

square-based pyramid

parallelogram-based prism

hexagonal pyramid

hexagonal prism

8.3 Surface Area of a Triangular Prism

Student Text pp. 253–258

Key Ideas Review

Unscramble the letters to spell the missing words.

- The surface _____ of a triangular _____ is the sum of the areas
       AAER   IMPRS

  of its _____.
         ACEFS

Example: Calculate Surface Area

Calculate the surface area of the triangular prism.

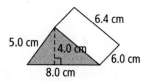

Solution

Method 1:
Find the area of the triangular base.

$A = \frac{1}{2} \times b \times h$

$= \frac{1}{2} \times 8.0 \times 4.0$

$= 16.0$

Find the areas of the rectangular sides.

Side 1 = 8.0 × 6.0
= 48.0

Side 2 = 5.0 × 6.0
= 30.0

Side 3 = 6.4 × 6.0
= 38.4

Find the sum of the three rectangular sides and two triangular bases.

16.0 + 16.0 + 48.0 + 30.0 + 38.4 = 148.4

The surface area is 148.4 cm².

Method 2:
Make a net of the triangular prism.
Fill in the area of each shape in the net.

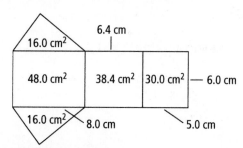

Add the areas together.

16.0 + 16.0 + 48.0 + 30.0 + 38.4 = 148.4

The surface area is 148.4 cm².

Name: _____   Date: _____

Practise

1. a) What is the area of the prism's triangular base?

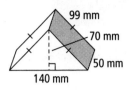

b) What are the areas of the rectangular faces?

c) Calculate the sum of two triangular faces and three rectangular faces. What is the surface area of the prism?

2. Calculate the surface area of each triangular prism.

a)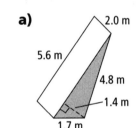

b)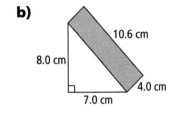

8.3 Surface Area of a Triangular Prism • MHR   75

8.4 Volume of a Triangular Prism

Student Text pp. 259–263

Key Ideas Review

Circle True or False.

1. The base of a triangular prism is a rectangle. — True   False

2. The volume of a triangular prism is equal to the area of all the triangular bases multiplied by the height of the prism. — True   False

3. Volume of prism = Area of a triangular base × Height of prism — True   False

Example: Find Volume

Find the volume of cheese.

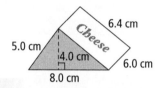

Hint

Area of a triangle $= \frac{1}{2} \times$ Base $\times$ Height

Solution

The base of the triangular prism is the shaded triangle.

Volume = Area of base × Height

$V = \left(\frac{1}{2} \times 8.0 \times 4.0 \right) \times 6.0$

$V = 96$

The volume of cheese is 96 cm³.

Making Connections

You'll see triangular prisms at home as a door stopper, a wedge of cheese, and even as a skateboard ramp.

Find the dimensions of several triangular prisms at home, school, or at a local park or sports arena. Which type of triangle do you see most often for the base of the prism?

Practise

1. Calculate the volume of the triangular prism.

   a) What is the height of the prism?

   b) What is the area of the triangular base?

   c) Calculate the product. What is the volume of the prism?

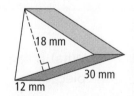

2. Calculate the volume of each triangular prism.

   a)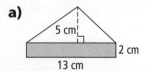

   b)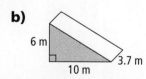

8.5 Surface Area or Volume of Triangular Prisms

Student Text pp. 264–267

Practise

1. The dimensions of a nylon tent are given.

   a) How much nylon is needed to make this tent?

   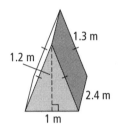

   b) What is the capacity of the tent?

2. The top 5 cm were cut off a triangular block of wood.

   a) What was the volume of the original triangular prism?

   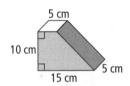

   b) What is the volume of the piece that was removed?

   Strategies
   Make a diagram or a picture of the original block of wood.

   c) What is the volume of the remaining piece?

   Making Connections

   Package design is an important part of selling a product.

   The shape of the container also influences the cost of packaging.

   Examine a rectangular prism. Divide the prism into two congruent triangular prisms. If you were planning to market a new type of cheese, which polyhedral container would you use? Consider the cost of wrapping for the container.

Chapter 8: Reviewing for the Test

8.1 Recognize and Sketch Three-Dimensional Figures

page 244

1. Which diagrams could not be a front, top, or side view of this three-dimensional figure?

A.   B.

C.   D.

E.

2. a) Which was the front view?

   b) Which was the top view?

   c) Which was the side view?

3. The top view of a prism is shown. What type of prism is it?

8.2 Build Models of Three-Dimensional Figures

page 248

4. Sketch the skeleton of each three-dimensional figure. How many edges and vertices does each figure have?

   a) triangular prism

   b) pentagonal pyramid

5. The numbers on the opposite faces of proper dice add to seven. Which of the following nets of cubes could be for dice?

A.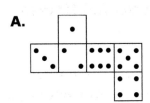

B.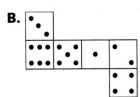

C.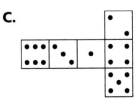

D.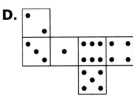

E.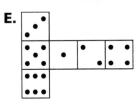

6. How many faces does a hexagonal pyramid have?

8.3 Surface Area of a Triangular Prism

page 253

7. Calculate the surface area of each triangular prism.

a)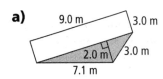

b)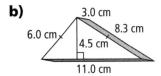

8. The surface area of a prism's triangular base is 4.6 m². If the prism has a height of 0.4 m and the triangle has a perimeter of 9.5 m, what is its total surface area?

Name: _____ Date: _____

8.4 Volume of a Triangular Prism

page 259

9. Calculate the volume of each triangular prism.

   a)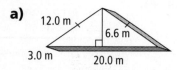

   b)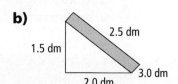

10. The surface area of a prism's triangular base is 1001 cm². If the prism has a height of 8 cm, what is its volume?

8.5 Surface Area or Volume of Triangular Prisms

page 264

11. a) The door and walls of a playhouse are to be painted and the floor is to be carpeted. It costs $2.00 per square metre for paint and $8.50 per square metre for carpet. What is the total cost?

    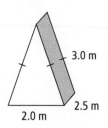

    b) What is the house's capacity?

12. The volume of a wedge of cheese is 714 cm³. The surface area of its base is 51 cm². What is the height of the wedge?

Name: _____ Date: _____

9.1 Collect, Organize, and Use Data
Student Text pp. 280–285

Key Ideas Review

- A **sample** is a small group taken from a population.
  A sample is often surveyed to make predictions about the whole group.

- A **census** is a survey in which everyone participates.

Label the population and the sample.

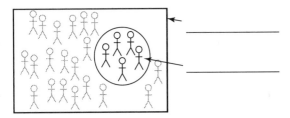

population
- the entire group of people you want to learn about

sample
- a small group that represents a population

Example: Use a Circle Graph to Display Data

At a big concert, one merchandise booth made $1000: $150 selling posters, $250 selling CDs, and $600 selling clothing.

a) Draw a circle graph to show the data.
b) All the booths made $4500. About how much money was made on CDs?

Literacy Connections

A sample is a small group from the population.

The sample should reflect the same characteristics as the population.

A sample is used to make predictions about the whole group or population.

Solution

a) Use a table to find the section angles.

Item	Sales ($)	Fraction	Decimal	Section Angle
Posters	150	$\frac{150}{1000} = \frac{3}{20}$	$3 \div 20 = 0.15$	$0.15 \times 360° = 54°$
CDs	250	$\frac{250}{1000} = \frac{1}{4}$	$1 \div 4 = 0.25$	$0.25 \times 360° = 90°$
Clothing	600	$\frac{600}{1000} = \frac{3}{5}$	$3 \div 5 = 0.6$	$0.6 \times 360° = 216°$

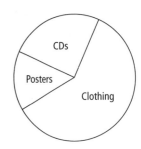

Draw a circle.
Use a protractor to measure each section.

b) CD sales were $\frac{1}{4}$ = 25% of the booth's total sales.

25% of the population is $0.25 \times \$4500 = \$1125$.
About $1125 was made selling CDs.

Practise

1. Winnerton School is hosting a cross-country meet. To plan the food, the principal surveyed 25 of her grade 8 students about which of four pizzas they would prefer.

   frequency
   - the number of times a particular piece of data or number appears in the data set

   a) Complete the frequency chart.

Topping	Tally	Frequency
All meat	||||| |||||		
Hawaiian	||||| |||	
Veggie	||||	
3 Cheeses	||	

   b) Express each as a fraction and a percent.

   c) Draw a circle graph to show the data.

2. The principal is expecting 200 grade 8 students at the cross-country meet.

   a) Did the principal choose an appropriate sample for this population?

   b) Use percents to predict how many students would prefer the Hawaiian pizza.

   c) Each student will receive 2 slices of pizza. Each pizza has 8 slices. How many Hawaiian pizzas should be ordered?

82   MHR • Chapter 9: Data Management: Collection and Display

Name: _____ Date: _____

9.2 Comparative Bar Graphs

Student Text pp. 286–291

Example: Read a Comparative Bar Graph

The students in two art classes were asked to choose which of sculpting, painting, or drawing was their favourite activity. Ms. Kwan has 30 students and Mr. Nguyen has 32 students. The results were converted into percents and graphed using a comparative bar graph.

a) Which class has a higher percentage who like drawing?
b) How many students like sculpting best?

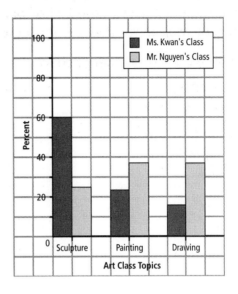

Solution

a) The drawing bar for Mr. Nguyen's class (37.5%) is taller than the drawing bar for Ms Kwan's class (16.6%), so Mr. Nguyen's class has a higher percentage of people who like drawing.

b) 60% of Ms. Kwan's class prefers sculpting: $0.60 \times 30 = 18$.
25% of Mr. Nguyen's class prefers sculpting: $0.25 \times 32 = 8$.
$18 + 8 = 26$ students said they like sculpting best.

comparative bar graph
- a bar graph in which two or more groups of data are shown side by side

Practise

1. Each year, Ms. Diamond has her students keep track of the average weekly mass of their classroom's garbage and recycling. Draw a comparative bar graph for the information.

Year	2001	2002	2003	2004
Garbage (kg)	14.2	12.5	8.8	6.3
Recycling (kg)	5.4	6.8	8.1	10.2

9.2 Comparative Bar Graphs • MHR 83

9.3 Histograms

Student Text pp. 292–297

Key Ideas Review

Observe the differences and similarities between the two graphs.

Bar Graph

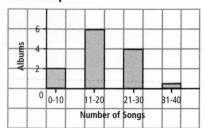

Histogram

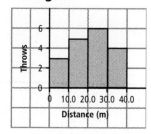

histogram
- a connected bar graph that shows data organized into intervals

- Equal-sized intervals
- Data are integers
- Bars separated

- Equal-sized intervals
- Data can be any decimal number
- Bars touching

Literacy Connections

For the bar graph, the height of each bar indicates the number of albums containing a number of songs in a particular range. The heights of the bars add up to 13, so 13 albums were examined in total. The first bar (0–10) indicates that two albums contained up to 10 songs each, the second bar indicates that six albums contained between 11 and 20 songs, and so on.

The histogram can be read in the same way.

Practise

1. Everyone in a grade 8 class ran 200 m. The students' times are given, in seconds.

40.4	41.8	37.0	30.7	45.2	35.4	40.8
36.4	34.1	41.8	30.9	28.7	40.7	37.9
42.6	35.4	38.0	35.0	45.6	34.9	37.8
35.3	31.3	43.0	33.7	41.5	37.1	36.7

a) Organize the data in the tally chart.

Time (s)	Tally	Frequency
25.0–29.9		
30.0–34.9		
35.0–39.9		
40.0–44.9		
45.0–49.9		

b) Draw a histogram to display the data.

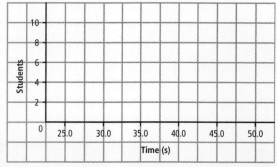

84 MHR • Chapter 9: Data Management: Collection and Display

9.4 Use Databases to Solve Problems
9.5 Use a Spreadsheet to Present Data and Solve Problems

Student Text pp. 298–309

Key Ideas Review

Match the missing words with their locations in the spreadsheet.

	A	B	C	D	E
1	select	graph	computer	organized	spreadsheet
2	bubbles	database	electronic	data	entry
3	sort	comparison	information	popcorn	graphs
4	answer	sandals	numeric	solve	event

1. A _____ (E1) is a software tool used to organize and display _____ (C4) data; it can be used to develop types of _____ (E3).

2. A _____ (B2) is an _____ (D1) collection of information.

3. _____ (C2) databases allow you to select specific information and _____ (A3) the information in different ways.

Example: Choosing a Graph

Max, Gina, and Jacob work together. Last week, Max worked 16.5 hours, Gina worked 17.5 hours, and Jacob worked 16 hours. What type of graph should be used to display this data?

Solution

The data is not appropriate for a line graph or a histogram, so choose either a bar graph or a circle graph.

The numbers are all really close together, so it may be difficult to see the difference on a circle graph.

Choose a bar graph.

Practise

1. In each situation, would you use a database or a spreadsheet?

    a) Sorting students alphabetically by last name

    b) Graphing assignment marks

    c) Calculating interest in several bank accounts

Name: _____ Date: _____

2. Volunteer information is kept in a database.

Last Name	First Name	Age	Driver	Hours per Week
Boggan	Tom	14	No	5
Patel	Tessa	16	No	8
Lanier	Yvette	23	Yes	12
Potter	Chad	28	No	12
Sachu	Mira	19	Yes	14

a) What information does the fourth column contain?

b) How are these records sorted?

3. The social committee used a spreadsheet to display how much money was spent on the school dance.

	A	B
1	Item	Cost ($)
2	Food	650
3	Decorations	275
4	Advertising	75
5	DJ	200

a) How can you use the spreadsheet to find the total amount spent?

Making Connections

Before an election, surveys or polls are often carried out to determine what the people want or are thinking.

Try conducting your own poll.

With a study partner, create a list of five questions on favourite hobbies or foods.

Survey at least five males and five females. The sample can be your friends, classmates, or family members.

Organize your data in a table or chart.

Graph the data using a pie chart or comparative bar graph.

Then write a short report about your findings.

b) What type of graph should be used to display this data?

Chapter 9: Reviewing for the Test

9.1 Collect, Organize, and Use Data

page 280

1. Match each sample group with the population from which it is taken.

   Sample

   a) 40 grade 8 girls are surveyed about using computers.

   b) People outside a grocery store are asked, "What is your favourite soup?"

   c) A group of people aged over 18 are asked whom they might vote for.

   Population

   A. all grade 8 girls

   B. all voters

   C. all shoppers at grocery stores

2. A company is testing four new flavours of pop. Researchers asked 24 people which they preferred.

Flavour	Frequency	Fraction
A	6	$\frac{1}{4}$
B	4	$\frac{1}{6}$
C	8	$\frac{1}{3}$
D	6	$\frac{1}{4}$

   Show this data in a circle graph.

   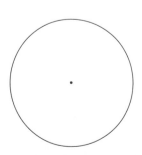

9.2 Comparative Bar Graphs

page 286

3. 13 boys and 17 girls were surveyed about their number of siblings. Convert to percents to compare the results.

Number of Siblings	Girls	Boys
More than 3	6	3
1 or 2	8	6
0	3	4

4. A treasurer kept track of the expenses and revenue of a fair.

   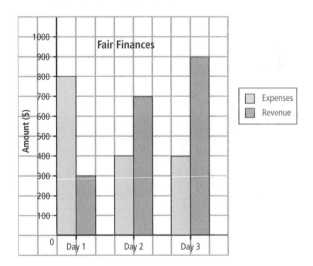

   a) Was money made or lost? Explain.

   b) About how much?

9.3 Histograms

page 292

5. A zoo graphed data about the lengths of the snakes in the reptile house.

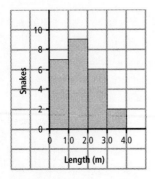

a) Which interval contains the most snakes?

b) How many snakes are at least 2 m long?

6. Antonina collected data about the ages of people at a family reunion.

36	21	18	7	52	55
8	11	58	13	39	41
2	16	16	87	14	53
83	79	55	56	24	41
38	10	4	34	25	27

Organize the data in a tally chart.

Age	Tally	Age	Tally
0–9		50–59	
10–19		60–69	
20–29		70–79	
30–39		80–89	
40–49		90–99	

9.4 Use Databases to Solve Problems
9.5 Use a Spreadsheet to Present Data and Solve Problems

page 298

7. In each situation, would you use a database or a spreadsheet?

a) Preparing a circle graph

b) Comparing hair and eye colour

c) Listing all students with food allergies

8. Rosa and Charlie compared marks on their history tests using a spreadsheet.

	A	B	C
1	Test	Rosa	Charlie
2	1	64%	77%
3	2	70%	73%
4	3	70%	68%
5	4	82%	84%

a) What is in cell B4? Explain.

b) What type of graph would best display this information? Explain.

c) Who is doing better? Explain.

d) Who is improving more? Explain.

10.1 Analyse Data and Make Inferences

Student Text pp. 318–323

Key Ideas Review

- You can analyse sets of data using various displays.
  - Line graphs can be used to analyse trends.
  - A stem-and-leaf plot can show the least and greatest values.
- You can make inferences based on data analysis.

Example 1: Analyse a Trend

Analyse the line graph for the temperature at various times on a day in winter.

a) What was the temperature at 9 a.m.?

b) When will the temperature be 2°C?

c) Predict when the temperature will be 5°C.

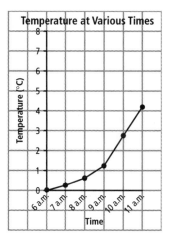

Solution

a) Locate the time 9 a.m. on the horizontal axis. Go straight up to the graph. Then go straight left to read the temperature on the vertical axis. The temperature was about 1.25°C.

b) Locate 2°C on the vertical axis. Go straight right to the graph. Then go straight down to the horizontal axis to read the year. In around 9:30 a.m., the temperature will be 2°C.

c) Extend the graph to predict a trend. This trend suggests that around 11:30 a.m., the temperature will be 5°C.

Example 2: Analyse a Set of Data

Several schools in the town of Goodwill collected donations for the local children's hospital. Jenna recorded the data in a stem-and-leaf plot. Describe the data. How much can the hospital expect from a school?

Donations ($)

Stem (thousands)	Leaf (hundreds)
10	0 4
11	5 5 7 8
12	0 1

Solution

Method 1: Analyse the Range of Data
The least donation is $10 000.

The greatest donation is $12 000.

The hospital can expect a donation between $10 000 and $12 000 from a school.

Donations ($)

Stem (thousands)	Leaf (hundreds)
10	0 4
11	5 5 7 8
12	0 1

range
- the range of the data is the difference between the least and greatest values in the set of data

Method 2: Calculate the Measure of Central Tendency
To get a better estimate, find the mean, median, and mode.

$$\text{Mean} = \frac{\text{sum of values}}{\text{number of values}}$$

$= (10\,000 + 10\,400 + 11\,500 + 11\,500 + 11\,700 + 11\,800 + 12\,000 + 12\,100) \div 8$

$= 91\,000 \div 8$

$= 11\,375$

The mean donation is $11 375.

The median is the middle value; in this data, the median is between $11 500 and $11 700, or $11 600.

The mode is the most frequently occurring value; in this data, the mode is $11 500.

Donations ($)

Stem (thousands)	Leaf (hundreds)
10	0 4
11	5 5 7 8
12	0 1

median

Practise

1. The temperatures as you climb a mountain are shown.

   a) Describe the temperature trend.

   b) Estimate the temperature at 1000 m.

   c) When did the temperature reach 9°C?

   d) Predict the temperature at 3000 m.

2. The swim team recorded the length of time each member could hold his or her breath.

   a) What is the shortest breath time?

   b) What is the longest breath time?

   c) What is the mean, median, and mode?

Breath Times (s)

Stem (tens)	Leaf (ones)
2	3 5 9
3	1 4 6
4	0 3 4 5 6 6 8

Name: _____ Date: _____

10.2 Understand and Apply Measures of Central Tendency

Student Text pp. 324–327

Key Ideas Review

- A measure of central tendency is a value around which a data set tends to be centred.

The following are measures of central tendency.

- Mean = $\dfrac{\text{sum of values}}{\text{number of values}}$
- **Median**: the middle value when the set of data is arranged in order from least to greatest
- **Mode**: the most common value in the set of data; sometimes there can be more than one mode, or none

- When a small set of data has a very high or very low value:
  - The mean is not a good measure of central tendency.
  - The median is not a good measure of central tendency.
  - The mode may not exist.
- Removing an unusual value from a small set of data can cause the mean to change.

Examine the data set of 100-m race times.

14.6 s 14.7 s 14.7 s 14.8 s 14.8 s
10.9 s 14.0 s 14.1 s 14.9 s 14.9 s
15.0 s 15.0 s 15.1 s 15.2 s 15.5 s

1. Select the mean of the race times.

   a) 14.9 s   b) 14.5 s   c) 14.1 s   d) 15.0 s

2. Select the median of the race times.

   a) 14.7 s   b) 14.8 s   c) 14.9 s   d) 15.0 s

3. Select the mode(s).

   a) 14.7 s   b) 14.8 s   c) 14.9 s   d) 15.0 s

4. Select the mean when the lowest value has been removed from the data.

   a) 14.5 s   b) 14.6 s   c) 14.7 s   d) 14.8 s

Practise

1. Find the mean, median, and mode for each data set.

   a) 11 16 17 16 18

   mean =

   median =

   mode =

   b) 55 70 85 65 60

   mean =

   median =

   mode =

Apply

2. Yan wrote 7 math tests. His scores in percents are shown.

   78 90 85 79 91 92 87

   a) Find the mean, median, and mode.

   mean =   median =   mode =

   b) Which measures of central tendency best describe Yan's performance? Explain.

   c) Why is the other measure of central tendency not a good choice? Explain.

10.3 Bias in Samples

Student Text pp. 328–333

Key Ideas Review

Circle the correct word for each statement.

1. An unbiased sample is **randomly** / **not randomly** drawn.

2. An unbiased sample **does not** / **does** reflect the characteristics of the population from which it is drawn.

3. A sample may be biased if it is too **small** / **large**.

Example 1: Biased Sample

Robert and David were asked to determine the measures of central tendency for the numbers of goals scored by their hockey team.

 4 5 3 2 1
 5 5 6 6 7

Robert decided to work with the first five values of the set of data (or a sample).

David decided to work with all of the values (or the population).

a) Find the mean, median, and mode for Robert's data set and for David's data set.
b) What is wrong with Robert's sample?

population
- the entire group of people you want to learn about

random sample
- a sample in which everyone in a population has an equal chance of being selected

Solution

a) **Robert's Sample**

 Mean = $\dfrac{\text{sum of values}}{\text{number of values}}$

 = (4 + 5 + 3 + 2 + 1) ÷ 5
 = 15 ÷ 5
 = 3

The mean number of goals is 3.

To find the median, order the values from least to greatest. Then find the middle value.

 1 2 3 4 5

The median number of goals is 3.

There is no mode.

David's Population

$$\text{Mean} = \frac{\text{sum of values}}{\text{number of values}}$$
$$= (4 + 5 + 3 + 2 + 1 + 5 + 5 + 6 + 6 + 6 + 7) \div 11$$
$$= 50 \div 11$$
$$\doteq 4.55$$

The mean number of goals is about 5.

To find the median, order the values from least to greatest. Then find the middle value.

    1 2 3 4 5 5 5 6 6 6 7

The median number of goals is 5.

There are two modes: 5 and 6.

b) Compare each method.

Measure of Central Tendency	Sample (Robert)	Population (David)
mean	3	5
median	3	5
mode	none	5 and 6

The results for Robert's sample were lower than the results for David's population. Robert's sample was not a good sample because three scores were very low compared to the others; the sample did not represent the population well.

Example 2: Random Sample and Sample Characteristics

Rose and Dana were asked to survey the grade 8 students in their school on their favourite school event. There are 30 students in grade 8.

Rose surveyed the first 15 grade 8 students to arrive at school.

Dana randomly surveyed student volunteers at the school bake sale.

a) What is wrong with each person's sample?
b) Describe a better sampling method.

Solution

a) Rose's sample contains bias. She surveyed half of the grade 8 students, those who entered the classroom first.
Dana's sample contains bias. Although she randomly surveyed students, she only surveyed volunteers at the bake sale. Not all students volunteer their time.
Neither sample represents the entire grade 8 student population at the school.

b) Students should be chosen at random to be surveyed. This could be done by selecting, say, every third student who enters the school.

Practise

1. Does each sample contain bias? Explain why or why not.

   a) A sales clerk asks two customers about the prices during the one-day sales event held at his store.

   b) The school guidance department surveyed five grade 8 students to determine the number of hours grade 8 students spend on volunteer work per week.
   Here are the results: 10 h   8 h   7 h   0 h   5 h

2. Is each sample randomly selected? Explain.

   a) Students in a class are assigned a number from 1 to 6.
   A number cube with numerals 1 to 6 is rolled seven times.
   Students whose assigned number appears are surveyed.

   b) Lars wants to determine if fast food should be served in the school cafeteria.
   He surveys students who have purchased french fries for lunch.

Apply

3. Mohan thinks that more vegetarian items should be offered in the school cafeteria. He surveys his friends who are vegetarian.

   a) Is this a random sample? Explain why or why not.

   b) Identify the bias in the sample.

   c) Describe how Mohan could conduct an unbiased survey.

10.4 Make and Evaluate Arguments Based on Data

Student Text pp. 334–339

Key Ideas Review

- It is important to evaluate claims that are based on data.
- Graphs may be distorted to support claims.
- Data can be used to build convincing arguments.

Example 1: Evaluate an Argument Based on Data

Sam is in charge of ordering the cake for a party on the last day of school. He surveys the students to determine which of three types is the most popular. Sam claims the majority of students prefer chocolate.

Cake	Tally	Frequency
chocolate	||||| ||||	9
vanilla	||||| ||	7
yellow	||||	4

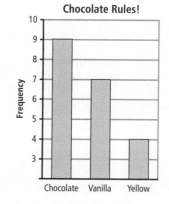

How has Sam distorted the data to make his argument?

Solution

The vertical scale on the graph is distorted.

It should start at 0. So if you stacked the bars for vanilla and yellow, the new bar will be taller than the bar for chocolate.

The title contains bias. It should just say what the graph is about, that is, Popular Cake Types.

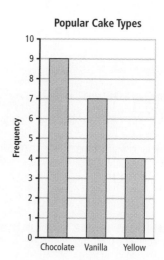

Name: _____ Date: _____

Example 2: Make an Argument Based on Data

Parkside School must choose between two students to represent the school in the provincial science competition.

The science test marks for the two top students are shown.

Student	Test 1	Test 2	Test 3	Test 4	Test 5
Mei	95%	85%	87%	86%	88%
Rory	80%	87%	90%	90%	91%

Make an argument, based on the data, that supports each student.

Solution

Argument to Support Mei

Mei's mean = (95 + 85 + 87 + 86 + 88) ÷ 5   Rory's mean = (80 + 87 + 90 + 90 + 91) ÷ 5
           = 441 ÷ 5   = 438 ÷ 5
           = 88.2   = 87.6

Mei's mean test score is higher than Rory's, so Mei should represent the school at the science competition.

Argument to Support Rory

Look at the performance trends for each student.

Rory's performance has continued to improve after the first test.

Mei's performance has not been consistent.

If you eliminate the lowest test score from Rory's data, his mean test score is 89.5.

If you eliminate the highest test score from Mei's data, her mean test score is 86.5.

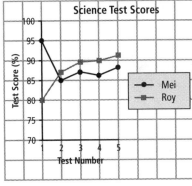

If the trends are extended it is expected that Rory will perform better than Mei on the next test. Therefore, Rory should represent the school at the science competition.

> **Making Connections**
>
> Find an advertisement or article that uses data to make a misleading statement.
>
> Identify:
> - what data are quoted
> - the statement that the data are used to support
> - why the data are misleading
>
> Write a brief summary of your findings.

Name: _____   Date: _____

Practise

1. The revenues from ticket sales for two movies on one day are shown.

   An entertainment magazine claims that *Forest Friends* was twice as popular as *Playground Pals*.

   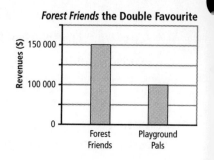

   a) How has the data been distorted to support the magazine's claims?

   b) Draw an undistorted bar graph for this problem.

   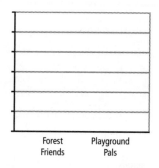

2. Nadia and Juan are being considered for a math medal.

   Their scores on the last five math tests are shown.

   Nadia: 90% 95% 92% 100% 88%
   Juan: 85% 90% 92% 100% 100%

   a) Calculate each person's mean test score. Based on the mean score, who should receive the math medal? Explain.

   b) The math medal will be given to the person with the best performance at the end of the school year. Predict who should receive the math medal. Draw a graph to help you explain.

   c) Explain and support your recommendation from part b) another way.

98 MHR • Chapter 10: Data Analysis: Analysis and Interpretation

Chapter 10: Reviewing for the Test

10.1 Analyse Data and Make Inferences

page 318

1. A music store tracks its sales to determine trends in popularity.

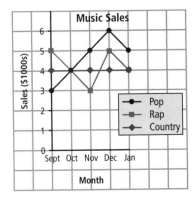

   a) Which category of music had the greatest sales, in which month, and what was the value of the sales?

   b) In which month were all music categories' sales the same?

   c) Predict the sales for each category in February.

2. Geometry project scores in percents for a class are shown.

   Describe five things that the set of data tells you.

Stem (tens)	Leaf (ones)
4	9
5	7 9
6	5 5 8 9
7	3 3 3 3 5 5 8
8	4 5 5 8
9	0 5

10.2 Understand and Apply Measures of Central Tendency

page 324

3. Sales for two electronics salespeople over the same week are shown.

Day	John's Sales ($)	Ron's Sales ($)
Monday	4000	2000
Tuesday	4000	5000
Wednesday	2000	3000
Thursday	3000	3000
Friday	4000	5000

   a) Calculate the mean, median, and mode for each salesperson.

Measure of Central Tendency	John	Ron
mean		
median		
mode		

   b) Based on the mean, who is the better salesperson?

   c) Based on the median, who is the better salesperson?

   d) Based on the mode, who is the better salesperson?

   e) Which measure of central tendency is the most important for identifying the better salesperson? Explain.

10.3 Bias in Samples

page 328

4. Tom and Tania surveyed grade 8 students on the number of hours spent on volunteer work.

   Tom asked six members of the volleyball team. His results are:

   1 h   0 h   5 h   7 h   5 h   6 h

   Tania asked every third student entering math class. Her results are:

   5 h   6 h   7 h   8 h   5 h   6 h
   7 h   6 h

   a) Calculate the mean, median, and mode for each sample.

Measure of Central Tendency	John	Ron
mean		
median		
mode		

   b) Determine whether or not each sample contains bias. Explain why or why not.

   c) Is each sample randomly selected? Explain.

   d) If a sample from part c) is not randomly selected, describe a better sampling method.

10.4 Make and Evaluate Arguments Based on Data

page 334

5. A scholarship is awarded to the student who shows the most promise for future studies. Recent grade averages, in percents, are shown for the two finalists.

Student	Grade Average (%)		
	Term 1	Term 2	Term 3
Tony	87	90	88
Emma	84	88	92

   Use measures of central tendency and graphs to help you answer parts a) and b).

   a) Make a convincing argument to support why Tony should win.

   b) Make a convincing argument to support why Emma should win.

11.1 Add Integers

Student Text pp. 350–355

Key Ideas Review

- If two integers in a sum have the same sign, add the two numerals, and keep the sign the same.
- If two integers in a sum have opposite signs, the result depends on the relative sizes of the integers.

opposite integers
- two integers with the same numeral but different signs
- 5 and –5 are opposite integers

zero principle
- the sum of opposite integers is zero
- 3 + (–3) is an example

1. Use the two facts to complete the chart.

	Integer Sum	Positive or Negative?
a)	(+1) + (+9) =	
b)	(+6) + (–8) =	
c)	(–2) + (–5) =	
d)	(–7) + (+8) =	

Example: Add Integers

Find each sum in three ways:

- using integer chips
- using a number line
- using mental math

a) 5 + (+3)
b) –3 + (–2)
c) –2 + 2
d) 6 + (–3)

Hint

In this workbook,
⊖ represents –1.
⊕ represents +1.

Solution

a) **Method 1: Using Integer Chips**

5 + (+3) = 8

Method 2: Using a Number Line
The integers have the same sign, so add the numerals and keep the sign.

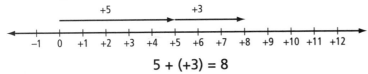

5 + (+3) = 8

Method 3: Using Mental Math
5 + (+3) = (+5) + (+3)
 = +8
 = 8

Hint

A positive number can be written without the positive sign.

b) Method 1: Using Integer Chips

   −3 + (−2) = −5

Method 2: Using a Number Line
The integers have the same sign, so add the numerals and keep the sign.

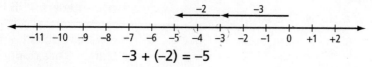

−3 + (−2) = −5

Method 3: Using Mental Math
−3 + (−2) = −5

c) Method 1: Using Integer Chips

   −2 + (+2) = 0

Method 2: Using a Number Line

−2 + (+2) = 0

Method 3: Using Mental Math
The integers have the same numeral, but different signs. Apply the zero principle.
−2 + (+2) = 0

d) Method 1: Using Integer Chips

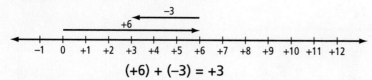   (+6) + (−3) = +3

Method 2: Using a Number Line

(+6) + (−3) = +3

Method 3: Using Mental Math
The integers have opposite signs but different numerals. 6 > 3 and 6 is positive so the answer will be positive. Subtract the numerals.
(+6) + (−3) = +3

Practise

1. What integer sum is shown? Give each result.

   a)

   b)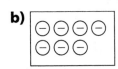

   c)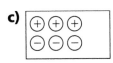

   d)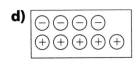

2. What integer sum is shown? Give each result.

   a)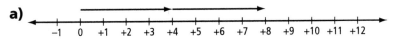

   b)

   c)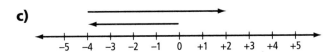

3. Use integer chips to model each sum. Give each result.

   a) 4 + (−10) =

   b) −5 + (−6) =

4. Use a number line to model each sum. Give each result.

   a) −3 + (−3) =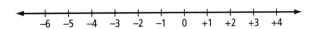

   b) −6 + 6 =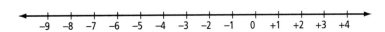

Name: _____ Date: _____

5. Decide whether each sum is positive, negative, or zero. Give each result.

Integer Sum	Positive, Negative, Zero	Result
2 + (−5)		
7 + (−7)		
−8 + (−2)		
−9 + 11		

6. Add.

   a) 12 + 7 =

   b) 35 + (−23) =

   c) −40 + (−51) =

   d) −20 + 48 =

   e) −21 + 14 + (−19) =

   f) 8 + (−14) + (−11) + (−2) =

Apply

7. The table shows the performance of two stocks over five days in one week. CompU started the week at $19. CompY started the week at $16. Which company ended the week with the higher price?

Stock	Monday	Tuesday	Wednesday	Thursday	Friday
CompU	−2	+3	+4	−2	+1
CompY	+1	−1	+5	−3	+6

Name: _____   Date: _____

11.2 Subtract Integers
Student Text pp. 356–360

Key Ideas Review

Fill in the blanks with the words, diagrams, or numbers from the list.
One number may be used more than once.

| addition | ⊖⊖⊖ | ⊖ | ⊕ | ⊕⊕⊕ | (–3) |
| opposite | ⊖⊖⊖⊖⊖ | –5 | + | –2 | +3 |

1. To show –2 – 3 with **integer chips**,
   • add _____ and _____ to ⊖⊖.
   • then take away _____.
   • the result is _____.

2. To show –2 – 3 with a **number line**,
   • start at _____, then "jump" over to _____.
   • The result is _____. It is the difference and direction
     between the two integers.
   • Complete the number line.

   ⟵ –9 –8 –7 –6 –5 –4 –3 –2 –1 0 +1 +2 +3 +4 ⟶

3. You can express a subtraction as the _____ of the _____ integer.
   –2 – 3 = –2 ___ ___
   = _____

Example 1: Subtract Integers

Show each subtraction by
• modelling with integer chips
• using a number line
• adding the opposite integer

a) 5 – (–2)   b) –3 – (–2)

Solution

a) Using Integer Chips

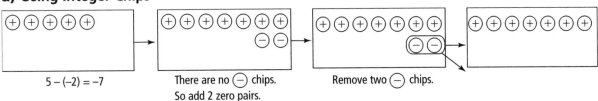

5 – (–2) = –7   There are no ⊖ chips.   Remove two ⊖ chips.
                  So add 2 zero pairs.

Using a Number Line
From –2 to 5 is 7 steps to the right.

⟵ –3 –2 –1 0 +1 +2 +3 +4 +5 +6 +7 +8 +9 +10 ⟶

11.2 Subtract Integers • MHR   105

Adding the Opposite Integer

$5 - (-2) = 5 + (+2)$ Change the subtraction operator to addition. Change
$ = 5 + 2$ the sign of the second integer to the opposite.
$ = 7$

b) Using Integer Chips

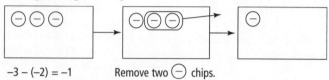

$-3 - (-2) = -1$ Remove two ⊖ chips.

Using a Number Line

From −2 to −3 is 1 step to the left.

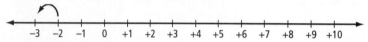

Add the Opposite Integer

$-3 - (-2) = -3 + (+2)$ Change the subtraction operator to addition. Change
$ = -3 + 2$ the sign of the second integer to the opposite.
$ = -1$

Example 2: Use Subtraction to Solve Problems

Express each of the following using integer subtraction. Then evaluate and interpret the result.

a) Tommy borrowed $4, then he borrowed $1.
b) Janice had 3 points, then lost 6 points.

Solution

a) Borrowed $4 means −4, borrowed $1 means −1.
 −4 − 1 is the integer subtraction expression.

Using Integer Chips

−4 − 1 can be rewritten as −4 − (+1).

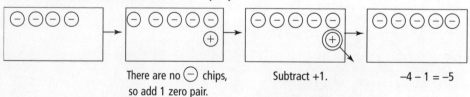

There are no ⊖ chips,   Subtract +1. $-4 - 1 = -5$
so add 1 zero pair.

Using a Number Line

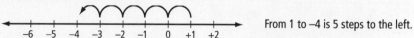

From 1 to −4 is 5 steps to the left.

Add the Opposite Integer

$-4 - 1 = -4 - (+1)$
$ = -4 + (-1)$ Change the subtraction operator to addition. Change
$ = -5$ the sign of the second integer to the opposite.

b) Had three points means +3, lost six points means –6.
3 – 6 is the integer subtraction expression.

Using Integer Chips

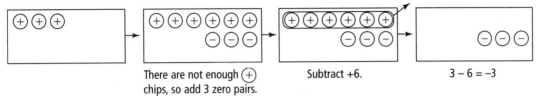

Using a Number Line

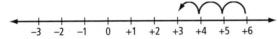

From 6 to 3 is 3 steps to the left.

Add the Opposite Integer

3 – 6 = 3 – (+6)
      = 3 + (–6)   Change the subtraction operator to addition.
      = –3   Change the sign of the second integer to the opposite.

Practise

1. Match each integer subtraction with its integer chip model. Then give the result.

  a) 3 – 4   **b)** –4 – (–7)   **c)** –4 – (–2)   **d)** 2 – (–4)

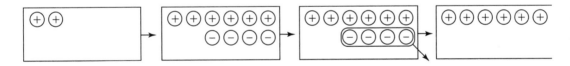

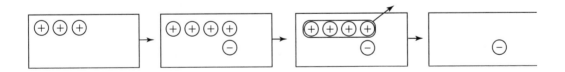

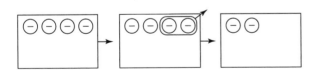

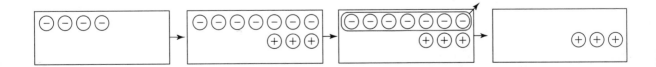

Name: _____   Date: _____

2. Model each integer subtraction using a number line. Then give the result.

   a) 2 – 5 =
   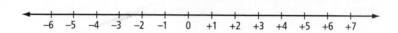

   b) –2 – 5 =
   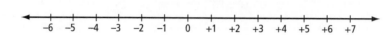

   c) –1 – (–5) =
   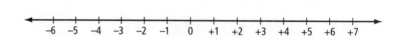

   d) 1 – (–5) =
   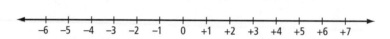

3. For each integer subtraction, add the opposite integer. Then give the result.

   a) 7 – 4 =   b) –8 – 6 =   c) –9 – (–12) =   d) 5 – (–7) =

4. Which integer statements give the same result? Which method did you use to check?

   a) 5 – 11   b) 5 + (–11)   c) 5 – (–11)
   d) –5 + 11   e) 11 – 5

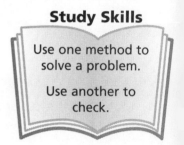

Study Skills

Use one method to solve a problem.

Use another to check.

5. Fill in the boxes with the correct integers.

   a) 7 – ☐ = –10   b) –1 – ☐ = –4   c) 8 – ☐ = 9

   d) ☐ – 11 = –3   e) ☐ – 4 = –6   f) ☐ – (–3) = –6

Apply

6. Express each problem using integer subtraction. Then evaluate and interpret the results.

   a) Raj earned $10, then bought two books for $4 with the money.

   b) The high temperature on Monday was –3°C. The low temperature was –10°C.

   c) Josh lost 9 points then gained 7 points.

   d) Mai owed $11, then repaid $5.

108   MHR • Chapter 11: Integers

11.3 Multiply Integers

Student Text pp. 361–365

Key Ideas Review

1. Examine the pattern in the sign chart for multiplication. Circle the correct word to complete each sentence.

×	+	−
+	+	−
−	−	+

   When multiplying two integers,
   a) if the signs of the two integers are the same, the product is **positive / negative**.

   b) if the signs are different, the product is **positive / negative**.

2. Circle True or False for each integer equation.

   a) $2 \times (-1) = 1 \times (-2)$ True   False

   b) $-3 \times (-4) = 4 \times 3$ True   False

Example 1: Apply Integer Multiplication

Write each as an integer multiplication. State the result and its meaning.

a) Each month for six months, Maria withdrew $10 from her piggy bank to make a donation to the animal protection society. How much did she withdraw in total?

b) Four sisters each borrowed $5 from their aunt. Their aunt decided to cancel the debts. How much did their aunt forgive?

Solution

a) $6 \times (-10) = -60$

   A total of $60 was withdrawn.

   A withdrawal is $10. That's −10.
   There are six months, that's $6 \times (-10)$.
   That's −$60.

b) $-4 \times (-5) = 20$

   One debt is $5. That's −5.
   Cancelling four debts is −4.
   There are four cancelled debts of $5.

   A total of $20 is forgiven.

   That's $20.

Example 2: Multiply More Than Two Integers

Multiply.

a) $2 \times (-4) \times (-3)$
b) $-4 \times (-3) \times (-2) \times 2$

11.3 Multiply Integers • MHR **109**

Solution

a) $2 \times (-4) \times (-3)$
   $= -8 \times (-3)$ There are two integers.
   $= 24$ The signs are the same.
   The product is positive.

b) $-4 \times (-3) \times (-2) \times 2$
   $= 12 \times (-2) \times 2$
   $= -24 \times 2$ There are two integers.
   $= -48$ The signs are different.
   The product is negative.

Practise

1. Multiply.

   a) $7 \times 9 =$
   b) $6 \times (-3) =$
   c) $-11 \times 3 =$
   d) $-10 \times (-2) =$

   e) $8 \times (-15) =$
   f) $9 \times 9 =$
   g) $-4 \times (-7) =$
   h) $-6 \times 15 =$

2. Multiply.

   a) $2 \times (-1) \times 3 =$
   b) $-4 \times (-2) \times (-4) =$

   c) $8 \times 9 \times (-2) \times (-2) =$
   d) $7 \times (-3) \times (-2) \times (-1) \times (-4) =$

Apply

3. Write each as an integer multiplication expression. State the result and give its meaning.

   a) At 11 a.m., the temperature outside is increasing by 3°C/h. What is the temperature change after 5 h?

   b) At 7 p.m., the temperature outside is decreasing by 2°C/h. What is the temperature change after 4 h?

   c) Candi withdrew $15 from the bank machine on four different days. Her dad deposited the money back into her account. How much was deposited?

Study Skills

Explain to a study partner how to determine the product of several integers. If your partner understands, then you most certainly do too!

4. a) Fill the square with five −1s and four +1s so the product of each row, diagonal, and column is negative.

   b) Fill the square with five +1s and four −1s so the product of each row, diagonal, and column is positive.

11.4 Divide Integers

Student Text pp. 366–369

Key Ideas Review

1. Examine the pattern in the sign chart for division. Circle the correct word to complete each sentence.
   When dividing two integers,

   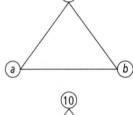

   a) if the signs of the two integers are the same, the quotient is **positive** / negative.

   b) if the signs are different, the quotient is positive / **negative**.

2. Examine the triangle that shows the following:

   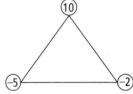

Multiplication	Related Division
$c = a \times b$	$c \div a = b$ or $\dfrac{c}{a} = b$
$c = b \times a$	$c \div b = a$ or $\dfrac{c}{b} = a$

   Circle the correct multiplication and division integer statements for the integer triangle.

   a) $5 \times 2 = 10$ b) $-5 \div 10 = 2$
   c) $-5 \times (-2) = 10$ d) $10 \div (-5) = -2$
   e) $-2 \times (-5) = 10$ f) $10 \div (-2) = -5$

Example: Divide Integers to Solve Problems

Write each as an integer division. State the result and its meaning.

a) The inside temperature of a freezer must be lowered by 15°C. The temperature decreases by 3°C per hour. How many hours will it take?

b) Avril borrowed $60 from her sister. She will make twelve equal monthly payments. What will be the monthly payment?

Solution

a) Divide the total temperature change by the size of the change per hour.
   $-15 \div (-3) = 5$
   It will take 5 h for the freezer temperature to decrease by 15°C.

b) Divide the amount of the loan by the number of months.
   $-60 \div 12 = -5$
   The monthly payment will be $5.

Name: _____   Date: _____

Practise

1. Draw a number triangle for each multiplication. Then write the related division statements.

a) 7 × (−3) = −21

b) −8 × 2 = −16

c) −4 × (−7) = 28

d) 12 × (−11) = −132

2. Divide.

a) $\dfrac{10}{5} =$

b) $\dfrac{-9}{-3} =$

c) $\dfrac{64}{-16} =$

d) $\dfrac{-121}{-11} =$

3. Find each quotient.

a) −16 ÷ 4 =

b) −27 ÷ 9 =

c) 28 ÷ (−7) =

d) −25 ÷ (−5) =

e) −45 ÷ (−5) =

f) −110 ÷ 5 =

g) −144 ÷ (−12) =

h) 56 ÷ 8 =

Apply

4. Write an integer division expression. Give the result and its meaning.

Making Connections

You learned how to calculate the mean in section 11.1.

a) A stocked decreased by $30 over 5 days. What was the mean daily decrease in price?

b) A scuba diver rose a total of 35 m in 7 stages. What was the mean rise per stage?

c) The temperature decreased by 12°C from 7 p.m. to 10 p.m. What was the mean hourly temperature decrease?

11.5 Order of Operations With Integers

Student Text pp. 370–373

Key Ideas Review

1. Select the correct solution for $4 - 3(2^2 - 10)$.

   a) $4 - 3(2^2 - 10)$
   $= 1(2^2 - 10)$
   $= 1(4 - 10)$
   $= 1(-6)$
   $= -6$

   b) $4 - 3(2^2 - 10)$
   $= 4 - 3(4 - 10)$
   $= 4 - 3(-6)$
   $= 1(-6)$
   $= -6$

   c) $4 - 3(2^2 - 10)$
   $= 4 - 3(4 - 10)$
   $= 4 - 3(-6)$
   $= 4 + 18$
   $= 22$

Hint

Order of Operations
Brackets
Exponents
Division and
Multiplication from left to right
Addition and
Subtraction from left to right

Example: Golf Scores

Honi played 9 holes of golf. She had the following scores:

$-2, -2, -2, -2, +1, 0, -1, 0, -1$

What was her mean score?

Solution

Method 1: Use Addition

$\text{Mean} = \dfrac{-2 + (-2) + (-2) + (-2) + (+1) + 0 + (-1) + 0 + (-1)}{9}$

$= \dfrac{-8 + (+1) + 0 + (-1) + 0 + (-1)}{9}$ Zeros don't affect the mean, so they can be ignored.

$= \dfrac{-7 + (-1) + (-1)}{9}$

$= \dfrac{-8 + (-1)}{9}$

$= \dfrac{-9}{9}$

$= -1$

Hint

To find the mean, divide the sum of the scores by the total number of scores.

Honi's average score was -1.

Method 2: Use Multiplication and Addition

$\text{Mean} = \dfrac{-2 + (-2) + (-2) + (-2) + (+1) + 0 + (-1) + 0 + (-1)}{9}$

$= \dfrac{4 \times (-2) + (+1) + 2 \times (-1)}{9}$ There are four -2s and two negative -1s. The zeros can be ignored.

$= \dfrac{-8 + (+1) + (-2)}{9}$

$= \dfrac{-7 + (-2)}{9}$

$= \dfrac{-9}{9}$

$= -1$

Honi's average score was -1.

Practise

1. Evaluate. Check using a calculator. Each answer represents a colour. Use the clues to colour the picture.

   a) $(21 - 5) \div 4$ —Red

   b) $3 + 4^2 \times 2$ —Orange

   c) $(4 + 5) \times 2$ —Black

   d) $-20 \div (-4) - 3$ —Brown

   e) $-2 \times 6 + (-3)^2$ —Blue

   f) $77 \div 7 \times 3$ —White

   g) $5 \times [-2 - (-3)]$ —Green

   h) $6 \times (7 - 3) + 5 \times 7 - 2$ —Pink

   i) $-100 \div (7 - 2)^2$ —Yellow

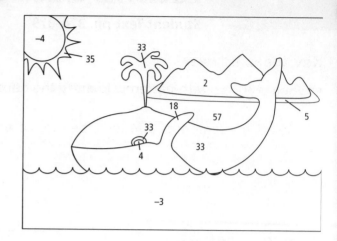

Technology Tip

Exponential expressions may be evaluated in these ways, using a calculator:

Example: 4^2

[C] 4 [x^2] [=]

[C] 4 [^] 2 [=]

[C] 4 [×] 4 [=]

Apply

2. Find the mean score for 7 holes of golf. Show two ways to find the answer.

   $+2, -1, -1, -1, -2, -3, -1$

3. You owe your sister $75. You babysit for $7.50/h to repay the loan.

   a) How much do you still owe after 4 h of work?

   b) How many more hours do you have to work to repay the loan in full?

11.6 Patterns and Trends With Integers

Student Text pp. 374–379

Key Ideas Review

- When working with patterns, identify the rules for the pattern.

1. Select the sentences that describe the pattern.

   −3, −6, −9, −12, −15, …

   a) The pattern starts at −3.
      The numbers decrease by 3 each time.
      The next three numbers in the pattern are: −18, −21, −24
      Each number in the pattern is divisible by 3 and −3.

   b) The pattern starts at −3.
      The next number is three times the one before.
      The next three numbers in the pattern are: −18, −21, −24
      Each number in the pattern is divisible by 3 and −3.

- To solve patterns with integers, look for patterns and trends.

2. Select the sentences that describe the pattern in the graph.

   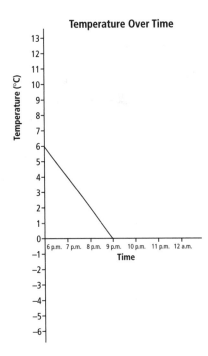

   a) The pattern starts at +6.
      The next temperature is $\frac{2}{3}$ the one before.
      Each number in the pattern is divisible by 2 or −2.
      By 12 a.m., the temperature will likely be −6°C.

   b) The pattern starts at +6.
      The next temperature decreases by 2°C from the one before.
      Each number in the pattern is divisible by 2 or −2.
      By 12 a.m., the temperature will likely be −6°C.

Name: _____   Date: _____

Practise

1. Match the pattern in Column A with its description in Column B.

   A
   a) 0, −2, −4, −6
   b) −8, −4, 0, 4
   c) −512, −256, −128, −64
   d) 7, −14, 28, −56

   B
   • The next number decreases by 2 from the one before.
   • The next number is $\frac{1}{2}$ times the one before.
   • The next number is −2 times the one before.
   • The next number increases by 4 from the one before.

2. For each pattern in question 1, determine the next four numbers.

   a)

   b)

   c)

   d)

3. Which pattern(s) in question 1 has factors of 4 and −4? Explain how you know.

Apply

4. The change in the price of a stock in dollars over four days is shown.

Day	Change in Stock Price ($)
1	−6
2	−5
3	−3
4	−1

   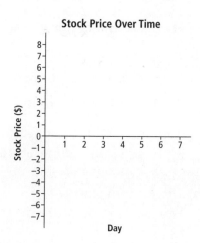
   Stock Price Over Time

   a) Graph the data.

   b) Predict change in the stock price on Day 7. Explain how you got your answer.

116   MHR • Chapter 11: Integers

Chapter 11: Reviewing for the Test

11.1 Add Integers

page 350

1. Add.

   a) 8 + 7 =

   b) –10 + 10 =

   c) –9 + (–9) =

   d) 6 + (–9) =

   e) –18 + 9 + (–5) =

   f) 7 + (–4) + (–3) + (–1) =

2. Cory's scores for six holes of golf are:

   –3, +1, –3, +1, –3, +1

   Determine his mean score.

11.2 Subtract Integers

page 356

3. Subtract.

   a) 12 – 17 =

   b) –9 – 11 =

   c) –10 – (–4) =

   d) –21 – (–9) – 7 =

11.3 Multiply Integers

page 361

4. Multiply.

   a) 8 × (–4) =   b) –2 × (–7) =

   c) 5 × (–12) =   d) 20 × (–3) =

   e) –2 × (–4) × (–1) =

   f) 2 × (–1) × 3 × (–5) =

11.4 Divide Integers

page 366

5. Divide.

   a) 77 ÷ (–11) =   b) –56 ÷ (–8) =

   c) –81 ÷ 9 =   d) –121 ÷ (–11) =

6. For each situation, write an integer division expression. Evaluate the expression and state its meaning.

   a) A scuba diver dived 30 m in six stages. What was the mean depth change of each stage?

   b) The temperature dropped 18°C at a constant rate over 6 h. What was the hourly temperature change?

11.5 Order of Operations With Integers

page 370

7. Evaluate without using a calculator. Check using a calculator.

   a) $5 \times 6 + 7 - 8$

   b) $4 - 6 + 7 \times 4 \div 2$

   c) $3 - 6(4 + 2)$

   d) $2(4^2 - 5)$

   e) $12^2 - 11^2 + 7$

   f) $-18 \div 6 + 2(-3 + 5)$

8. Place brackets in each equation to make the equation true.

   a) $3 + 4 \times 2 = 14$

   b) $28 \div 7 - 3 \times (-4) = -28$

11.6 Patterns and Trends With Integers

page 374

9. Determine the next four numbers in each pattern. Explain how you got your answer.

   a) 7, 4, 1, –2, …

   b) –8, 16, –32, 64, …

10. The temperature outside was 10°C at 8 a.m. Use the graph to determine the temperature at 2 p.m. Explain how you got your answer.

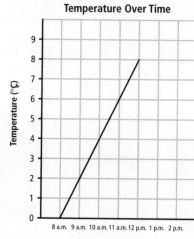

Temperature Over Time

12.1 Model and Solve Equations

Student Text pp. 388–393

Key Ideas Review

- Equations can be solved by inspection, using a model, or performing the opposite operation on both sides of the equal sign.
- To check solutions, substitute your answer into the equation and solve. The left side should equal the right side.

Practise

1. Each cup contains the same number of counters.

   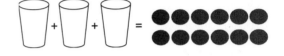

   a) What equation is modelled by the diagram?

   b) What operation needs to be undone to get 1 cup alone? What is its opposite operation?

   c) How many counters are in each cup?

2. Solve the equation modelled by each diagram.

   a)

   b)

3. Model each equation with a diagram. Solve the equation.

   a) $2k = 10$

   b) $4 - y = 1$

12.2 Apply the Opposite Operations

Student Text pp. 394–399

Key Ideas Review

Fill in the blanks with the words from the list.

> addition   subtraction   multiplication   division

- _____ and _____ are opposite operations.
- _____ and _____ are opposite operations.
- When undoing the operations performed on the variable, follow the reverse of the order of operations.

$9 \times b - 6 = 48$
undo first → $9 \times b$
undo second → -6

Example: Model and Solve an Equation

Gael is selling woven blankets at a craft sale. Half were bought on the first day and 8 more on the second day. There was 1 blanket unsold.

a) Write an equation to model Gael's blankets.
b) Solve the equation by applying the opposite operations.
c) Verify your solution.

Solution

a) Let b be the number of blankets at the beginning.
Half were sold on the first day, so I'll write that as $b \div 2$.
8 more were sold on the second day, so I'll write that as $b \div 2 - 8$.
There was only 1 at the end, so I'll write that as $b \div 2 - 8 = 1$.

b) $b \div 2 - 8 = 1$
$b \div 2 - 8 + 8 = 1 + 8$ Add to undo subtraction.
$b \div 2 = 9$
$b \div 2 \times 2 = 9 \times 2$ Multiply to undo division.
$b = 18$

c) Left Side $= b \div 2 - 8$ Right Side $= 1$
$= (18) \div 2 - 8$
$= 9 - 8$
$= 1$
Left Side = Right Side

Practise

1. For each equation, identify the operations that need to be undone in order.

    a) $56 = 2r - 6$

    b) $j \div 2 - 2 = 11$

2. Each cup contains the same number of counters.

    a) What equation is modelled by the diagram?

    b) What new equation do you get when you add 7 to both sides?

    c) How many counters are in each cup?

    d) Check your solution.

3. Solve the equation modelled by each diagram. Check your solution.

    a)

    b)

4. Model and solve each equation. Verify your solution.

    a) $3a + 4 = 10$

    b) $20 - 6y = 2$

5. Natalie bought 4 equal bags of peaches. After sharing 6 peaches, she was left with 14 for herself.

    a) Write an equation to model Natalie's peaches.

    b) How many peaches were in each bag?

12.3 Model Problems With Equations

Student Text pp. 400–404

Practise

1. Look at the pattern of marbles.

   Diagram 1   Diagram 2   Diagram 3

   a) Complete the table.

Diagram	Number of marbles	Pattern
1	4 = 1 + 3	1 + 3 × 1
2		
3		
4		

   b) Write an equation that models the pattern.

   c) How many marbles are in diagram 5?

   d) What diagram in the pattern uses 43 marbles?

2. Mail-order CDs cost $12 and there is a $4 fee for postage and handling. What is the price of ordering one CD? two CDs? ten CDs? Use an equation.

3. The inside edges of this ornament are 1.4 cm shorter than the outside edges. The perimeter of the inner pentagon is 22.8 cm. What is the perimeter of the outer pentagon?

   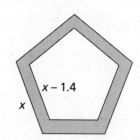

12.4 Explore Inequalities

Student Text pp. 405–411

Key Ideas Review

- An inequality uses a symbol to compare numbers or expressions. It can involve a variable.
- Match the inequality symbols with the possible wordings.

    >   at most
    ≥   more than
    <   no less than
    ≤   under

Example 1: Find the Solution Set

a) Write the model as an inequality using ≤ or ≥.

b) Rewrite the inequality using < or >.

c) Write the whole number solution set.

solution set
- a list of all the numbers that make a mathematical statement true

Solution

a) Let n represent the unknown number.
Since every number is at least 6, I write $n \geq 6$.

b) Since every number is over (or more than) 5, I write $n > 5$.

c) The solution set is every number from 6 on.
$n = 6, 7, 8, 9, 10, \ldots$

Example 2: Model Inequalities

Marilyn has 7 loonies in her coin bank and at least 2 more in her change purse. Model this situation with an inequality.

Solution

Let L represent the number of loonies.

At least 2 more loonies means the number is greater than or equal to 2.

If there are 7 more loonies, then $L \geq 2 + 7$ or $L \geq 9$.

Practise

1. Write an inequality that models each situation.

    a) That shirt costs at most $25.

    b) More than 20 000 people came.

2. Write each model as an inequality using < or >.

    a)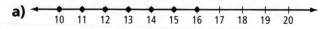

    b)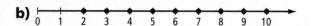

3. Rewrite the inequalities in question 2 using ≤ or ≥.

    a)

    b)

4. Write the whole number solution set for each inequality.

    a) $m + 1 > 7$

    b) $3n \leq 12$

    c) $4 > p$

    d) $q \geq 1001$

5. Use the clues to figure out each secret number.

    a) $x > 22$ $x \leq 25$ x is a multiple of 3

    b) $y \geq 32$ $y < 40$ y is a prime number

    c) $z \geq 90$ $z \leq 110$ z is a perfect square

Chapter 12: Reviewing for the Test

12.1 Model and Solve Equations

page 388

1. **a)** Each box contains the same number of balls. Which expression models this diagram?

   A $7 - 3p = 5$ **B** $3 - 7B = 5$

   C $3x - 7 = 5$ **D** $7b - 3 = 5$

   b) How many balls are there in each box?

2. Solve each equation. Verify your solution.

   a) $6k = 72$

   b) $h - 34 = 26$

   c) $3 + y = 101$

12.2 Apply the Opposite Operations

page 394

3. Solve the equation modelled by this diagram. Check your solution.

4. Solve each equation.

   a) $4 + 6w = 40$

   b) $x \times 5 - 8 = 27$

   c) $39 = 10m - 11$

   d) $n \div 2 + 3 = 17$

12.3 Model Problems with Equations

page 400

5. A library charges 50¢ to use a computer and 10¢ for every minute of Internet access.

   a) Write an equation to model this situation. Define your variables.

   b) How much would you pay for 1 hour and 40 minutes? Write your answer in dollars.

6. The outside edges of this ornament are 0.7 cm longer than the inside edges. The perimeter of the inner hexagon is 25.2 cm. What is the perimeter of the outer hexagon?

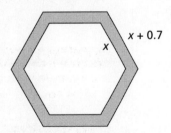

12.4 Explore Inequalities

page 405

7. a) Which inequality models this diagram?

$8 + 8 > \square$

   A $7 > b$ **B** $7 < b$

   C $b > 7$ **D** $8 \geq b$

   b) What is the whole number solution set?

8. Find the whole number solution set for each inequality.

   a) $7 + a \leq 9$

   b) $3c \geq 30$

   c) $2 < g - 11$

   d) $j + 1.5 \geq 12.5$

Name: _____ Date: _____

13.1 Internal Angles of a Triangle

Student Text pp. 424–429

Key Ideas Review

1. Use the diagrams to determine the sum of the angles in a triangle.

   a)   b)   c) 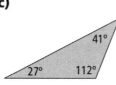   d)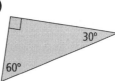

   a) 40° + 55° + 85° = _____   b) 55° + 55° + 70° = _____

   c) 27° + 112° + 41° = _____   d) 30° + 60° + 90° = _____

2. The measures of the internal (or interior) angles in a triangle add to _____.

3. Identify the type of triangle in each diagram of question 1.

   a)

   b)

   c)

   d)

> **scalene triangle**
> - a triangle with all interior (or inside) angles less than 90°
>
> **isosceles triangle**
> - a triangle with two sides equal
> - the base angles are congruent
>
> **right triangle**
> - a triangle containing a 90° angle
>
> **equilateral triangle**
> - a triangle with all sides equal
>
> **obtuse triangle**
> - a triangle with one angle that is between 90° and 180°

Practise

1. Find the measure of angle x in each triangle.

   a) 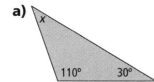   b)

   c) 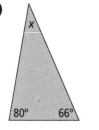   d)

   e) 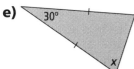   f)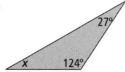

13.1 Internal Angles of a Triangle • MHR 127

Name: _____ Date: _____

13.2 Angle Properties of Intersecting and Perpendicular Lines

Student Text pp. 430–135

Key Ideas Review

- Two intersecting lines form pairs of opposite angles. Opposite angles are equal.
- Two angles whose measures add to 180° are supplementary angles.

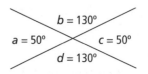

$a + b = 180°$, a and b are supplementary angles.
$c + d = 180°$, c and d are supplementary angles.

- Two angles whose measures add to 90° are complementary angles.

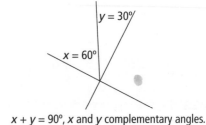

$x + y = 90°$, x and y complementary angles.

Identify the complementary angles. Use the letters to complete the riddle.

W: 90° and 90°   E: 25° and 75°   R: 30° and 60°
R: 115° and 65°   I: 39° and 51°   G: 6° and 84°
O: 20° and 160°   N: 123° and 57°   U: 149° and 11°
G: 85° and 95°   H: 45° and 45°   T: 27° and 63°

Why can you never win an argument with a 90°-angle?

Because they are always ___ ___ ___ ___ ___!

Example: Opposite, Supplementary, and Complementary Angles

a) a and z are opposite angles. What is a?
b) b and z are supplementary angles. What is b?
c) c and z are complementary angles. What is c?

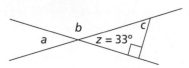

Solution

a) Opposite angles are equal.
   $a = z$
   $a = 33°$

b) The measures of supplementary angles add to 180°.
   $b + z = 180°$
   $b + 33° = 180°$
   $b - 33° = 180° - 33°$
   $b = 147°$

c) The measures of complementary angles add to 90°.

$c + z = 90°$
$c + 33° = 90°$
$c - 33° = 90° - 33°$
$c = 57°$

Practise

1. Find the measure of the angle that is complementary to the given angle.

   a) 70°   **b)** 2°   **c)** 45°

2. Find the measure of the angle that is supplementary to the given angle.

   a) 100°   **b)** 2°   **c)** 78°

3. Label the other three angles formed by these intersecting lines.

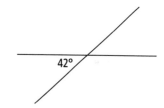

4. Find the measure of angle x.

   a) 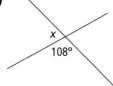   **b)**

   c)    **d)**

Name: _____   Date: _____

13.3 Angle Properties of Parallel Lines
Student Text pp. 436–441

Key Ideas Review

When a transversal intersected these parallel lines, eight angles were created.

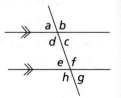

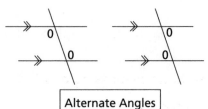

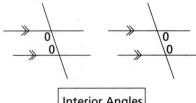

Complete the table to describe the angle relationships in the diagram at the top right.

Corresponding angles (equal)	Alternate angles (equal)	Interior angles (supplementary)
• a and e • • •	• c and e •	• c and f •

Example: Parallel Lines and a Transversal

∠JLK = 75°. What is the measure of ∠LOP?

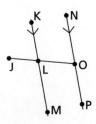

Solution

Method 1:
∠JLK and ∠LON are corresponding angles, so ∠JLK = ∠LON.
∠LON = 75°.
∠LON and ∠LOP are supplementary angles, so ∠LON + ∠LOP = 180°.
   ∠LOP = 180° − ∠LON
        = 180° − 75°
        = 105°

Method 2:
∠JLK and ∠KLO are supplementary angles, so ∠JLK + ∠KLO = 180°.
∠KLO = 180° − ∠JLK
     = 180° − 75°
     = 105°
∠KLO and ∠LOP are alternate angles, so ∠KLO = ∠LOP.
∠LOP = 105°

Practise

1. Use another method to find the measure of ∠LOP.

   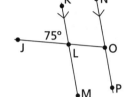

   a) How are ∠JLK and ∠MLO related?

   b) What is the measure of ∠MLO?

   c) How are ∠MLO and ∠LOP related?

   d) What is the measure of ∠LOP?

2. Find the measure of angle x in each diagram.

   a)

   b)

   c)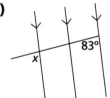

   d)

13.3 Angle Properties of Parallel Lines • MHR   131

Name: _____   Date: _____

3. Find the measures of the angles in each diagram. Give reasons.

a)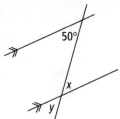

b)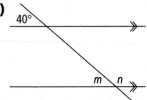

c)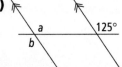

d)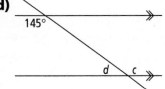

Making Connections

You can see parallel lines with an intersecting line everywhere.

You can see the lines in fences, in scaffolding, and in shelving.

Find a fence that has parallel lines that are intersected by a cross-piece.

Make a drawing of the fence. Measure the corresponding, alternate, and supplementary angles. Add these measures to your labels.

13.4 Apply Angle Measures

Student Text pp. 442–446

Practise

1. Find the measure of angle z.

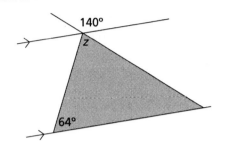

2. Two boats are 3 km away from a shore that runs north and south.
   Boat A is 25° north of west from Radio Station G.
   Boat B is 50° south of west from Radio Station G.

   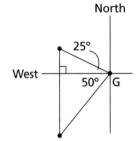

   a) What is the measure of ∠GAB?

   b) What is the measure of ∠GBA?

3. Two boats are both 5 km away from Radio Station G. Boat C is 15° north of west from the station and Boat D is 53° north of west from the station.

   a) Draw a diagram of this situation.

   b) What is the measure of ∠GCD?

   c) What is the measure of ∠GDC?

13.5 Construct Line Segments and Angles

Student Text pp. 447–451

Practise

1. Follow the steps to construct a triangle with side lengths of 5 cm, 5 cm, and 3 cm. A 5-cm line segment has been drawn for you.

   - Set your compass to 5 cm.
     Put the compass on one end of the segment.
     Draw part of a circle.

   - Set your compass to 3 cm.
     Put the compass on the other end of the segment. Draw part of a circle that intersects with the first one.

   - Draw two line segments to complete the triangle.

   5 cm

2. **a)** Construct a triangle with side lengths of 4 cm, 4 cm, and 4 cm. A 4-cm line segment has been drawn for you.

   4 cm

   b) What kind of triangle have you constructed?

   c) What would happen if you tried to make a triangle with side lengths of 4 cm, 2 cm, and 2 cm?

3. Suppose you were given an 80° angle.

   a) How could you use paper folding to construct a 40° angle?

   b) How could you use paper folding to construct a 60° angle?

Chapter 13: Reviewing for the Test

13.1 Internal Angles of a Triangle

page 424

1. Which of the following could be the angle measures of a triangle?

   A 55°, 55°, 90°

   B 24°, 48°, 108°

   C 23°, 63°, 83°

   D 50°, 57°, 73°

2. The measures of two angles of a triangle are given. What is the measure of the third angle?

   a) 20° and 70°

   b) 44° and 66°

   c) 35° and 110°

   d) 17° and 136°

13.2 Angle Properties of Intersecting and Perpendicular Lines

page 430

3. Find the measure of the angle that is supplementary to the given angle.

   a) 125°

   b) 34°

4. Find the measure of the angle that is complementary to the given angle.

   a) 81°

   b) 34°

5. What is the measure of each unknown angle?

13.3 Angle Properties of Parallel Lines

page 436

6. a) What is the measure of each unknown angle?

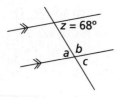

13.4 Apply Angle Measures

page 442

7. a) Find the measure of angle *m*.

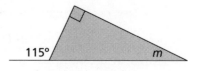

b) Find the measure of angle *r*.

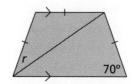

b) Which pair of angles are alternate angles?

   A *a* and *z*

   B *b* and *z*

   C *a* and *b*

   D *b* and *c*

13.5 Construct Line Segments and Angles

page 447

8. Suppose you are given a 100° angle. How could you use paper folding to construct a 25° angle?

c) Which pair of angles are corresponding angles?

   A *a* and *z*

   B *b* and *z*

   C *a* and *c*

   D *c* and *z*

d) Which pair of angles are interior angles?

   A *a* and *z*

   B *b* and *z*

   C *a* and *c*

   D *b* and *c*

Name: _____   Date: _____

Summer Tune-up: A Note to Parents/Guardians

This section of the workbook contains several practical and useful activities in the context of summer fun and outings. The focus is on review and practice of grade 8 concepts and skills in preparation for grade 9. The activities have been constructed so that they can be done anywhere: in the back seat of a car, on the beach, or under a shady tree. Several different kinds of puzzles are included. No special tools other than a pencil, paper, ruler, and calculator are required. Learning doesn't have to end with the school year!

If you have access to a computer, additional features are available at www.mcgrawhill.ca. Log on, and navigate to the grade 8 workbook page. Then, follow the links.

Name: _____ Date: _____

Getting Started

Start with this puzzle to get your brain thinking in mathematical terms.
This word search puzzle includes mathematical terms that you should be familiar with.
As you find each term, review the meaning of the term.

Mathematics Word Search

P	N	S	B	F	N	N	A	I	Z	C	X	G	U	D
A	O	I	A	R	E	A	B	Y	M	P	A	K	W	G
R	I	Z	X	A	Y	E	U	H	N	E	B	E	X	F
A	T	D	E	C	I	M	A	L	M	C	N	A	V	I
L	A	R	E	T	A	L	I	U	Q	E	E	R	H	X
L	M	A	Y	I	P	H	R	I	L	M	D	P	Z	P
E	I	N	X	O	E	U	W	A	U	H	M	I	V	O
L	T	U	S	N	R	S	C	L	N	C	G	V	A	D
O	S	E	L	E	C	S	O	S	I	E	B	C	K	N
G	E	P	R	R	E	V	K	J	L	N	A	Y	P	V
R	E	C	T	A	N	G	L	E	Y	B	A	H	P	B
A	N	B	W	U	T	X	E	G	M	F	L	U	C	I
M	L	Q	I	Q	E	I	C	F	H	J	Q	U	F	E
S	F	M	V	S	Q	R	O	K	V	F	A	R	V	I

Angle
Area
Decimal
Equilateral
Estimation
Fraction

Isosceles
Mean
Median
Parallelogram
Percent

Ratio
Rectangle
Scalene
Square
Volume

A Walk Around the Zoo

Anwar, Hilda, and Ferenc are visiting a zoo. At the entrance, they are given a map of the grounds, as shown.

The zoo enclosure is a square. A circular pathway leads around it, and a direct pathway leads from the entrance to the exit—a distance of 500 m.

1. The zoo charges an admission fee of $10.00 per person. Hilda has three 20% discount coupons. GST at 7% and PST at 8% is applied to the discounted price.

   a) Calculate the admission fee after the 20% reduction.

   b) Calculate the taxes applied to the admission fee.

   c) What is the total price of a ticket, after the discount is applied, and taxes are added?

   d) What is the total cost of admission for the group?

2. a) Plot a path on the map that will take the group past each animal enclosure from the entrance to the exit. They cannot pass any enclosure more than once. Number the stops along the path.

   b) Ferenc estimates that the total distance to be walked, from part a), is less than 2 km. Do you agree? Support your answer with your own estimate. Include your reasons.

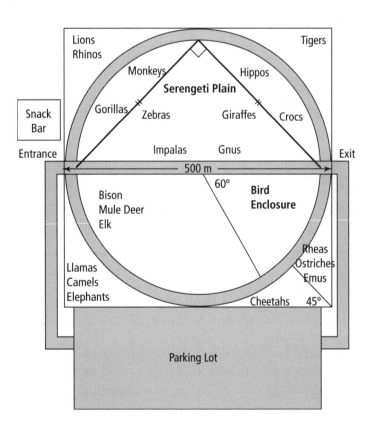

Name: _____ Date: _____

3. The bird enclosure contains 60 birds of various species. Zoo rules indicate that a minimum area of 500 m² must be allowed per bird. Does the enclosure satisfy the rule? Use calculations to support your answer.

4. The Serengeti Plain enclosure is an isosceles right triangle. Find the length of each of the equal sides of the triangle.

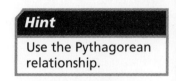

Hint

Use the Pythagorean relationship.

5. The Lion Shelter is built in the form of a triangular prism, as shown. The floor and roof are made of concrete. The ends are open.

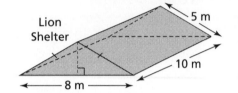

a) Calculate the surface area for the floor and roof.

b) The tallest lion in the zoo is 1.3 m high. Will she be able to walk upright through the shelter? Use calculations to support your answer.

c) Calculate the volume of the shelter.

6. A cheetah requires an area of 2500 m² to roam. How many cheetahs can the zoo keep in the enclosure provided?

7. Anwar's brother Hosni works at the snack bar. He receives a commission of 5% of the weekly sales. The sales for one week are shown in the table. Complete the table.

Item	Price per Item	Mon	Tues	Wed	Thurs	Fri	Sat	Sun	Total Sales	Total Revenue
Fruit juice	$1.50	24	32	26	18	28	38	42	$	
Hot dogs	$2.00	26	24	20	22	28	42	44	$	
Popcorn	$1.25	56	48	37	39	45	65	72	$	
Veggie snacks	$2.25	18	22	19	17	26	32	38	$	

a) Calculate the total revenue from all items for the week.

b) Calculate Hosni's gross commission for the week.

c) Find the median and mean numbers of hot dogs sold.

d) Why is the mean is higher than the median? Explain.

Crossword

ACROSS

1. lines that never meet
3. a chart that uses a circle to present data
6. notation for large or small numbers using powers of 10
7. the perimeter of a circle
9. an experiment that imitates a real-world process
11. it's calculated using the formula *I = Prt*

DOWN

1. lines that meet at a right angle
2. a problem that uses estimation to rapidly calculate a ballpark answer
4. measurement error that favours some answers over others
5. payment based on a percentage of sales
6. software for calculating relations among data
8. software for storing data such as a CD collection
10. a short form to help remember the order of operations

Name: _____   Date: _____

Fermi Goes to the Beach

Amanda, Suki, and Elzbieta decided to spend a day at secluded Luna Beach.

The beach forms a semicircle of sand over table rock, as shown.

The sand can be dug into for about 80 cm before hitting the rock.

The sand grains measure about 1 mm across, on the average.

The water of the bay is protected and shallow, with a smooth table rock bottom. The depth increases to a maximum of about 1.8 m before leaving the bay.

While floating on some inner tubes of diameter 120 cm that they had brought with them, the girls decided to play a Fermi game.

Each one came up with a Fermi problem for the others to solve, based on the beach environment.

Make reasonable assumptions and estimations to solve these Fermi problems.

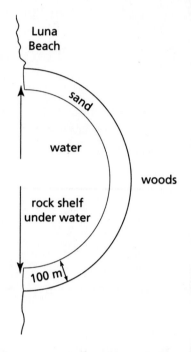

Name: _____ **Date:** _____

Amanda's Problem: How many grains of sand are there on this beach?

Name: _____   Date: _____

Suki's Problem: How many litres of water are there in this bay?

Elzbieta's Problem: How many floating inner tubes would fit on the surface of this bay?

Fraction/Decimal/Percent Match Game

Match each expression on the left with its corresponding answer on the right.

$\dfrac{13}{16} - \dfrac{5}{16}$ $1\dfrac{1}{2}$

$\dfrac{3}{8} + \dfrac{1}{4}$ $2\dfrac{3}{8}$

$\dfrac{3}{4} \times \dfrac{2}{5}$ $\dfrac{19}{20}$

$\dfrac{9}{14} \div \dfrac{3}{7}$ $\dfrac{3}{10}$

$1\dfrac{2}{5} + 2\dfrac{3}{10}$ 0.22

$3\dfrac{5}{8} - 1\dfrac{1}{4}$ $\dfrac{3}{8}$

$2\dfrac{4}{5}$ as a decimal $3\dfrac{7}{10}$

0.95 as a fraction $\dfrac{5}{9}$

22% as a decimal $\dfrac{1}{2}$

37.5% as a fraction $\dfrac{5}{8}$

2.8

Name: _____   Date: _____

Welcome to the Grade 9 BBQ!

Fridgeway High School welcomes its new grade 9 students with a BBQ day. Games are played and prizes are given out. In the Birthday Game, each student fills out a card with his or her name and birthday month. The cards are placed into a bin and mixed. Each contestant may select three cards at random. If at least two cards have the same birthday month, the contestant wins a prize.

1. Predict the probability of winning a prize. What is more likely: to win or not to win? Give a reason for your choice.

2. Simulate the game using 12 cards, one for each month. Use a deck of playing cards or create your own. Place the cards into a convenient container and mix. Select one card, and record the month in the table. Return the card, mix, and pick another. Repeat for a third card. Record whether you had a match or not.

   a) Play the game 20 times. Then, calculate the number of matches and non-matches.

   b) Use the results from the simulation to calculate the probability of winning the game.

   c) Did the probability from part b) agree with your prediction from Question 1?

Trial	Card 1	Card 2	Card 3	Match?
1				
2				
3				
4				
5				
6				
7				
8				
9				
10				
11				
12				
13				
14				
15				
16				
17				
18				
19				
20				
Total Matches				
Total Non Matches				

Preparing for Grade 9 • MHR 147

Integer Arranging

Fill in the blanks in the table according to the following rules:

a) you may select from the integers from −10 to −1 and 1 to 10.

b) the integer in the first column must be positive.

c) the integer in the second column must be negative.

d) you may use each integer only once in the puzzle.

_____ + _____ =

_____ − _____ =

_____ × _____ =

_____ ÷ _____ =

Checklist:

−10	−9	−8	−7	−6	−5	−4	−3	−2	−1
1	2	3	4	5	6	7	8	9	10

Cottage Angles

Mario's family has rented a cottage on a lake.

The cottage is an A-frame in the shape of an isosceles triangle.

The apex angle measures 50°. The second floor has an overhanging balcony that is parallel to the ground, forming a number of angles.

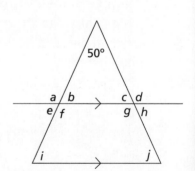

Calculate the measures of all of the angles.

a = ___   b = ___   c = ___   d = ___   e = ___

f = ___   g = ___   h = ___   i = ___   j = ___

148   MHR • Preparing for Grade 9

Telephone Scramble

In this puzzle, select one letter from the three in each box to spell out a word that describes an angle or pair of angles. The definitions are given below, but are not necessarily in the same order.

ABC 2	ABC 2	TUV 8	TUV 8	DEF 3							
ABC 2	MNO 6	MNO 6	PRS 7	JKL 5	DEF 3	MNO 6	DEF 3	MNO 6	TUV 8	ABC 2	PRS 7	WXY 9
MNO 6	ABC 2	TUV 8	TUV 8	PRS 7	DEF 3							
MNO 6	PRS 7	PRS 7	MNO 6	PRS 7	GHI 4	TUV 8	DEF 3					
PRS 7	GHI 4	GHI 4	GHI 4	TUV 8								
PRS 7	TUV 8	PRS 7	PRS 7	JKL 5	DEF 3	MNO 6	DEF 3	MNO 6	TUV 8	ABC 2	PRS 7	WXY 9

Definitions:

a) an angle whose measure is between 90° and 180°

b) two angles that add to 180°

c) the angle formed by two perpendicular lines

d) the angles formed by two intersecting lines

e) an angle whose measure is less than 90°

f) two angles that add to 90°

A Summer Business

Doris decided to see if she could start and run a lawn care service for the summer. She assumes that since most people go away on vacation during the summer, homeowners would need her services.

1. To check the market response, Doris created an advertising flyer on her computer. She offers a flat rate of $20 per lawn. She paid 6¢ each for 500 copies of the flyer, plus GST at 7% and PST at 8%. She then spent a day delivering flyers to 500 homes within a few blocks of her own house. How much has she invested in her market research?

2. Doris received 75 calls over the next two days. Determine the probability that a household will respond to one of her flyers.

3. **a)** Doris assumes that each customer will take a two-week vacation. Doris offers to mow lawns once a week for each week of vacation. If Doris has the whole summer booked for lawn mowing, write an equation that will determine the money, E, she will earn during the eight weeks of summer given a number of bookings, b.

   b) Doris now needs a lawnmower. She found a used self-propelled lawnmower in the newspaper for $200. She asked her father to lend her $200, and offered to repay the loan at the end of two months with simple interest at 5% per year. How much will Doris pay her father at the end of the two months?

c) Is this a safe investment for her father? Can he expect Doris to earn enough to pay back the loan? Explain and justify your answer.

d) Doris finds that the lawnmower burns 0.5 L of gasoline per hour of use. If gasoline costs 89¢/L, write an equation that will determine the gasoline costs, G, for the number of hours, h, of lawnmower use.

e) If Doris spends 10 hours per week mowing lawns, how much will she have earned over the summer, after gasoline expenses have been paid?

f) Summarize Doris's expenses and income for her summer business, as well as her net profit or loss. Use the chart.

Item	Expense	Earnings
Advertising		
Gasoline		
Lawnmower Loan		
Revenue		
Net Profit/Loss		

g) How much did Doris earn, on average, per hour?

h) Doris decided to cut each lawn for a flat rate of $20. Give two reasons why this might be a good idea, and two reasons why it is not.

Name: _____   Date: _____

Sierpinski: Not Just Triangles

You can produce a pattern called Sierpinski's Carpet by starting with 1 coloured square. For the second stage, start with a square of the same size, divide it into 9 squares, and colour the outer squares only.

 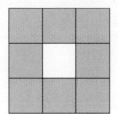 

The pattern continues by dividing each of the outer squares into 9 squares. In each of these, the outer 8 squares are coloured. Create this pattern in the blank square.

1. a) How many squares are coloured in the first stage? _____

   b) How many squares are coloured in the second stage? _____

   c) How many squares are coloured in the third stage? _____

2. Predict the number of coloured squares in the fourth stage: _____

3. Complete the table.

Stage	Number of Coloured Squares	Pattern
1		
2		
3		
n		

4. What is the rule for calculating the number of coloured squares in each additional stage?

5. Write the numbers of coloured squares for the first six stages of Sierpinski's Carpet as a sequence.

Breaking the Code

Letter substitutions are often used to code messages, such as:

NFFU BU FMFWFO UVFTEBZ

Code-breakers, known as cryptologists, try to break the code by determining the letters that occur most often in the English language, and then substituting these into the code, hoping to get enough of a clue to decode the message.

Consider the following paragraph:

"He walked out into the street and flagged down a taxi. The driver did not want to take him all the way to the airport, but an offer of a twenty-dollar tip convinced the driver otherwise. It was a trip of eighteen kilometres to the airport, which would take at least twenty minutes in the heavy traffic. He settled back in the seat, and listened to the news report on the radio. As the report ended, he could see a plane in the distance, climbing out after takeoff. Perhaps he might make his flight after all."

1. Make a tally chart of the occurrence of each letter in the paragraph.

Letter	Tally	Frequency	Letter	Tally	Frequency
A			N		
B			O		
C			P		
D			Q		
E			R		
F			S		
G			T		
H			U		
I			V		
J			W		
K			X		
L			Y		
M			Z		

2. Determine the six most frequently occurring letters in the paragraph.

3. Draw a bar graph showing the frequencies of the six most common letters.

4. The three most frequently used letters in the English language are E, T and A, in that order. Did your analysis of the paragraph give the same results? If not, explain why.

5. Represent each word in the coded message with an appropriate number of blanks. Find the most common letter in the message, and replace it with E. Replace the second most common with T, and the third with A. Can you decode the message?

___ ___ ___ ___   ___ ___   ___ ___ ___ ___ ___ ___   ___ ___ ___ ___ ___ ___ ___

6. Look for a pattern in the decoded message. What is the rule for letter substitution in the coded message?